MONOGRAPHS ON
STATISTICS AND APPLIED PROBABILITY

General Editors

**D.R. Cox, D.V. Hinkley, N. Reid, D.B. Rubin
and B.W. Silverman**

(Full details concerning this series are available from the Publisher)

The Analysis of Contingency Tables

SECOND EDITION

B.S. EVERITT

Professor of Statistics in Behavioural Science,
Institute of Psychiatry, London

CRC Press
Taylor & Francis Group
Boca Raton   London   New York

CRC Press is an imprint of the
Taylor & Francis Group, an **informa** business
A CHAPMAN & HALL BOOK

First published 1977 by Chapman & Hall

First edition 1977
Second edition 1992

Published 2019 by CRC Press
Taylor & Francis Group
6000 Broken Sound Parkway NW, Suite 300
Boca Raton, FL 33487-2742

© 1977, 1992 B.S. Everitt
CRC Press is an imprint of Taylor & Francis Group, an Informa business

First issued in paperback 2019

No claim to original U.S. Government works

ISBN 13: 978-0-367-45041-0 (pbk)
ISBN 13: 978-0-412-39850-6 (hbk)

Visit the Taylor & Francis Web site at
http://www.taylorandfrancis.com

and the CRC Press Web site at
http://www.crcpress.com

Library of Congress Cataloging-in-Publication Data

Catalog record is available from the Library of Congress.

Contents

Preface

The first edition of *The Analysis of Contingency Tables* arose from Professor A.E. Maxwell's earlier text *Analysing Qualitative Data* and there remain echoes of 'Max's' book in this second edition. I have, however, attempted to include more material on those methods which have developed over the last decade or so, for example logistic regression models for tables with ordered categories and for response variables with more than two categories. I have also given a brief account of that increasingly fashionable technique, correspondence analysis.

As in the first edition, the mathematical level of the book has been kept deliberately low in the hope that it will appeal to the same group of people as the original, namely research workers in psychiatry, the social sciences, psychology, etc., and students undergoing courses in medical and applied statistics.

My thanks are due to Dr Graham Dunn for his very helpful comments on the draft manuscript and to Ms Marie Dyer for her efficient typing.

B.S. Everitt
London

CHAPTER 1

Contingency tables and the chi-square test

1.1 Introduction

This book is primarily concerned with methods of analysis for **frequency data** occurring in the form of cross-classifications or **contingency tables**. In this chapter various terms will be defined, some necessary nomenclature introduced and a description given of how such data may arise. Later sections describe the chi-square distribution, and give a numerical example of testing for **independence** in a contingency table by means of the **chi-square test**.

1.2 Classification

It is possible to classify the members of a **population** – a generic term denoting any well defined class of people or things – in many different ways. People, for instance, may be classified into male and female, married and single, those who are eligible to vote and those who are not, and so on. These are examples of **dichotomous** classifications. **Multiple** classifications are also common, as when people are classified into left-handed, ambidextrous and right-handed, or, for the purpose of, say, a Gallup poll, into those who intend to vote (a) Conservative, (b) Labour, (c) Liberal Democrat, (d) those who have not yet made up their minds and (e) others. In this text interest will lie primarily in classifications whose categories are **exhaustive** and **mutually exclusive**. A classification is exhaustive when it provides sufficient categories to accommodate all members of the population. The categories are mutually exclusive when they are so defined that each member of the population can be correctly allocated to one, and only one, category. At first sight it might appear that the requirement that a classification be exhaustive is very restrictive. It might, for example, be of interest to carry out a Gallup poll, not on the voting intentions of the electorate as a whole, but only on those of university students. The difficulty is resolved if the definition of a population is recalled. The statistical definition of the word is more fluid than its definition in ordinary usage, so it is quite in order to define the population in question as 'all university students eligible to vote'. Categories too are adjustable and may often be altered or combined; for instance, in the voting example it is unlikely that much information would be lost by amalgamating categories (d) and (e).

When the population is classified into several categories it is possible to count the number of individuals in each category. These counts or frequencies are the type of data with which this book will be primarily concerned: in other words, with **qualitative data** rather than with **quantitative data** obtained from measurement of continuous variables such as height, temperature, and so on. In general, of course, information from only a **sample** from the whole population is available. Indeed, one main function of statistical science is to demonstrate how valid inferences about some population may be made from an examination of the information supplied by a sample. An essential step in this process is to ensure that the sample taken is a representative (unbiased) one. This can be achieved by drawing what is called a **random sample**, that is one in which each member of the population in question has an equal chance of being included. The concept of random sampling is discussed in more detail in Chapter 2.

1.3   Contingency tables

The main concern of this book will be the analysis of data which arise when a sample from some population is classified with respect to two or more qualitative variables. For example, Table 1.1 shows a sample of 5375 tuberculosis deaths classified with respect to two qualitative variables, namely sex and type of tuberculosis causing death. (Note that the categories of these variables as given in the table are both exhaustive and mutually exclusive.)

A table such as Table 1.1 is known as a **contingency table**, and this 2×2 example (the members of the sample having been dichotomized in two different ways) is its simplest form. Had the two variables possessed multiple rather than dichotomous categories the table would have had more cells than the four shown.

The entries in the cells for these data are frequencies. These may be transformed into proportions or percentages but it is important to note that, in whatever form they are presented, the data were

Table 1.1   *Deaths from tuberculosis*

	Males	Females	Total
Tuberculosis of respiratory system	3534	1319	4853
Other forms of tuberculosis	270	250	522
Tuberculosis (all forms)	3804	1571	5375

originally frequencies or counts rather than continuous measurements. Of course, continuous data can often be put into discrete form by the use of intervals on a continuous scale. Age, for instance, is a continuous variable, but if people are classified into different age groups, the intervals corresponding to these groups can be treated as if they were discrete units.

Since Table 1.1 involves only two variables it may be referred to as a **two-dimensional** contingency table; in later chapters we shall be concerned with three and higher dimensional contingency tables which arise when a sample is classified with respect to more than two qualitative variables.

1.4   Nomenclature

At this point it is convenient to introduce some general notation for two-dimensional tables. Later, when dealing with higher dimensional tables, this notation will be extended.

The general form of a two-dimensional contingency table is given in Table 1.2; here a sample of N observations is classified with respect to two qualitative variables, one having r categories and the other having c categories. It is known as an $r \times c$ contingency table.

Table 1.2   *General form of a two-dimensional contingency table*

		Columns (variable 2)					
		1	2 . . .			c	Total
Rows (variable 1)	1	n_{11}	n_{12} .	.	.	n_{1c}	$n_{1.}$
	2	n_{21}					$n_{2.}$
	r	n_{r1}				n_{rc}	$n_{r.}$
Total		$n_{.1}$	$n_{.2}$			$n_{.c}$	$n_{..} = N$

The observed frequency or count in the i category of the row variable and the j category of the column variable, that is the frequency in the ijth cell of the table, is represented by n_{ij}. The total number of observations in the ith category of the row variable is denoted by $n_{i.}$ and the total number of observations in the jth category of the column variable $n_{.j}$. These are known as **marginal totals** and in terms of the cell frequencies, n_{ij}, are given by:

$$n_{i.} = n_{i1} + n_{i2} + \cdots + n_{ic}$$

$$= \sum_{j=1}^{c} n_{ij}, \tag{1.1}$$

$$n_{.j} = n_{1j} + n_{2j} + \cdots + n_{rj}$$

$$= \sum_{i=1}^{r} n_{ij}. \tag{1.2}$$

Similarly

$$n_{..} = \sum_{i=1}^{r} \sum_{j=1}^{c} n_{ij} \tag{1.3}$$

$$= \sum_{i=1}^{r} n_{i.} = \sum_{j=1}^{c} n_{.j}. \tag{1.4}$$

$n_{..}$ represents the total number of observations in the sample and is usually denoted simply by N. This notation is generally known as **dot notation**, the dots indicating summation over particular subscripts.

In the case of the data shown in Table 1.1 we have:

(1) $r = c = 2$, that is both variables have two categories;
(2) $n_{11} = 3534$, $n_{12} = 1319$, $n_{21} = 270$ and $n_{22} = 252$ are the cell frequencies;
(3) $n_{1.} = 4853$ and $n_{2.} = 522$ are the row marginal totals, that is the total number of deaths from the two types of tuberculosis;
(4) $n_{.1} = 3804$ and $n_{.2} = 1571$ are the column marginal totals, that is the total number of males and females in the sample;
(5) $N = 5375$ is the total number of observations in the sample.

1.5   Independent classifications

Once the type of data of concern have been examined, it becomes possible to consider the questions which might be of interest. In general, the most important question is whether the qualitative variables forming the contingency table are independent or not. To answer this question, it is necessary to clarify just what independence between the classifications would entail. In the case of a 2 × 2 table this is relatively easy to see. For example, returning to the data of Table 1.1, it is clear that, if the form of tuberculosis from which people die is independent of their sex, the proportion of males who die from tuberculosis of the respiratory system would be expected to be equal to the proportion of females who died from the same cause. If these proportions differ, death from tuberculosis of the respiratory system tends to be **associated** more with one of the sexes than the other. (Of course, the two proportions might be expected to differ in some measure due solely to chance factors of sampling, and for other reasons which might be attributed to random causes; what needs to be ascertained is whether or not the observed difference between the proportions is too large to be attributed to such causes, and for this the test to be discussed in the next section is required.)

Having seen, intuitively, that independence in a 2 × 2 table implies the equality of two proportions, it is necessary to examine slightly more formally what the concept implies for the general $r \times c$ contingency table. First suppose that, in the population from which the sample is to be taken, the probability of an observation belonging to the ith category of the row variable **and** the jth category of the column variable is represented by p_{ij}; consequently the frequency F_{ij} to be expected in the ijth cell of the table resulting from sampling N individuals is given by:

$$F_{ij} = N p_{ij}. \tag{1.5}$$

(Readers familiar with mathematical expectation and probability distribution will recognize that $F_{ij} = E(n_{ij})$, under the assumption that the observed frequencies follow a **multinomial distribution** with probability values p_{ij}, see, for example, Mood and Graybill, 1963, Ch. 3.)

Now, let $p_{i.}$ represent the probability, in the population, of an observation belonging to the ith category of the row variable (in this case with no reference to the column variable), and let $p_{.j}$ represent the corresponding probability for the jth category of the column variable. Then from the multiplication law of probability, independence between the two variables, in the population, implies that:

$$p_{ij} = p_{i.} p_{.j}. \tag{1.6}$$

In terms of the frequencies to be expected in the contingency table independence is therefore seen to imply that:

$$F_{ij} = N p_{i.} p_{.j}. \qquad (1.7)$$

However, the reader might ask in what way this helps since the independence of the two variables has only been defined in terms of **unknown** population probability values. The answer is that these probabilities may in fact be **estimated** very simply from the observed frequencies, and it is easy to show that the 'best' estimates, $\hat{p}_{i.}$ and $\hat{p}_{.j}$, of the probabilities $p_{i.}$ and $p_{.j}$ are based upon the relevant marginal totals of observed values; that is:

$$\hat{p}_{i.} = \frac{n_{i.}}{N} \text{ and } \hat{p}_{.j} = \frac{n_{.j}}{N}. \qquad (1.8)$$

(These are **maximum likelihood estimates**, see Mood and Graybill, 1963, Ch. 12.) The use of the estimates of $p_{i.}$ and $p_{.j}$ given in equation (1.8) allows estimation of the frequency to be expected in the ijth cell of the table if the two variables are independent. Inspection of equation (1.7) shows that this estimate, which is generally represented as E_{ij}, is given by:

$$E_{ij} = N \hat{p}_{i.} \hat{p}_{.j}$$
$$= N \frac{n_{i.}}{N} \frac{n_{.j}}{N} = \frac{n_{i.} n_{.j}}{N}. \qquad (1.9)$$

When the two variables are independent, the frequencies estimated using formula (1.9) and the observed frequencies should differ by amounts attributable to chance factors only. If, however, the two variables are not independent larger differences would be expected to arise. Consequently it would seem sensible to base any test of the independence of the two variables forming a two-dimensional contingency table on the size of the differences between the two sets of frequencies n_{ij} and E_{ij}. Such a test is discussed in the following section. (In latter parts of this text the estimated expected frequencies E_{ij} will, when there is no danger of confusing them with the frequencies F_{ij}, often be referred to simply as 'expected values'.)

1.6   Chi-square test

In the preceding section the concept of the independence of two variables was discussed. To test for independence it was indicated that we need to investigate the truth of the hypothesis:

$$p_{ij} = p_{i.}p_{.j}. \tag{1.10}$$

In general this hypothesis will be referred to as the **null hypothesis**, denoted by the symbol, H_0.

It was also pointed out that the test should be based upon the differences between the estimated values of the frequencies to be expected when H_0 is true (that is the E_{ij}) and the observed frequencies (that is the n_{ij}). Such a test, first suggested by Pearson (1904), uses the statistic, X^2, given by:

$$X^2 = \sum_{i=1}^{r} \sum_{j=1}^{c} \frac{(n_{ij} - E_{ij})^2}{E_{ij}}. \tag{1.11}$$

It is seen that the magnitude of this statistic depends on the values of the differences $(n_{ij} - E_{ij})$. If the two variables are independent these differences will be less than would otherwise be the case; consequently X^2 will be smaller when H_0 is true than when it is false. Hence what is needed is a method for deciding on values of X^2 which should lead to acceptance of H_0 and those which should lead to its rejection. Such a method is based upon deriving a **probability distribution** for X^2 under the assumption that the hypothesis of independence is true. Acceptance or rejection of the hypothesis is then based upon the probability of the obtained X^2 value; values with 'low' probability lead to rejection of the hypothesis, others to its acceptance. This is the normal procedure for deriving **significance tests** in statistics. In general a 'low' probability is taken to be a value of 0.05 or 0.01, and is referred to as the **significance level** of the test. (For a more detailed discussion of significance level and related topics, see Mood and Graybill, 1963, Ch. 12.)

By assuming that the observed frequencies have a particular distribution, namely a multinomial distribution, and by further assuming that the expected frequencies are not too small (see page 39), the statistic X^2 may be shown to have approximately a **chi-square distribution**. The test of the hypothesis of independence may now be performed by comparing the calculated value of X^2 with the tabulated values of the chi-square distribution (see Appendix A), although in most cases this will not be necessary since the actual probability value associated with the calculated value of the test statistic under the null hypothesis will be available from whatever particular computer package is being used.

1.7 Chi-square distribution

Most readers will be familiar with the normal distribution, accounts of which are given in many statistical textbooks (see again Mood and Graybill, 1963, Ch. 6). The chi-square distribution arises from it as the probability distribution of the sums of squares of a number of independent variables, z_i, each of which has a standard normal distribution, that is one with mean zero and standard deviation unity. The form of the distribution depends upon the number of independent variates involved. For example, a chi-square variable, χ^2, formed by the sum of squares of v independent standard normal variables, namely:

$$\chi^2 = z_1^2 + z_2^2 + \cdots + z_v^2 \tag{1.12}$$

has a distribution depending only on v. Diagrams showing the different shapes the distribution takes for different values of v are given in many textbooks (for example, Hays, 1973, Ch. 11). In general the number of independent variates forming the chi-square variable is known as the **degree of freedom**. As will be seen in the next section, the degrees of freedom of the test statistic, X^2, depends upon the number of categories of each variable forming the table. Knowing this value, however, allows the relevant probability for judging whether or not the hypothesis of independence is compatible with the observations to be determined.

1.8 Degrees of freedom for a two-dimensional table

The statistic for testing the independence of the two variables forming a contingency table has already been seen to be:

$$X^2 = \sum_{i=1}^{r} \sum_{j=1}^{c} \frac{(n_{ij} - E_{ij})^2}{E_{ij}}. \tag{1.13}$$

The degrees of freedom of the chi-square distribution which approximates the distribution of X^2 when the hypothesis of independence is true, is simply the number of independent terms in (1.13), given that the row and column marginal totals of the table are fixed. The total number of terms in (1.13) is $r \times c$, that is the number of cells in the table. Some of these terms are, however, determined by knowledge of the row and column totals. For example, knowing the r row totals fixes r of the frequencies n_{ij}, one in each row, and hence determines r of the terms in (1.13). The number of independent terms in (1.13) is thus reduced to $(rc - r)$. If it is assumed that the frequency fixed by each row total is that in the last column, it can be seen that, of the c column totals, only the first $(c - 1)$ remain to be considered. Each of these fixes one frequency in the body of the table and consequently reduces the number of independent terms by one. Consequently $rc - r - (c - 1)$ independent terms are left in (1.13). This gives the degrees of freedom of X^2.

$$\text{d.f.} = rc - r - (c - 1)$$
$$= (r - 1)(c - 1). \quad (1.14)$$

(The degrees of freedom of a contingency table may also be regarded as the number of cells of the table which may be filled arbitrarily when the marginal totals are fixed.)

1.9 Numerical example

To help the reader understand the discussion so far, the data shown in Table 1.1 will be used. The hypothesis under test in this case is that the form of tuberculosis from which people die is independent of their sex – which is another way of saying that the two classifications are independent. The first step in the calculation of X^2 is to compute the expected values using formula (1.9). For example, E_{11} is given by:

$$E_{11} = \frac{4853 \times 3804}{5375}$$

$$= 3434.6 \quad (1.15)$$

The remainder of the calculation is set out in Table 1.3.

The differences $(n_{ij} - E_{ij})$ sum to zero, and the value of the test

Table 1.3 Calculating X^2 for the data of Table 1.1

(1) n_{ij}	(2) E_{ij}	(3) $(n_{ij} - E_{ij})$	(4) $(n_{ij} - E_{ij})^2$	(5) $(n_{ij} - E_{ij})^2/E_{ij}$
3534	3424.6	99.4	9880.36	2.88
1319	1418.4	− 99.4	9880.36	6.97
270	369.4	− 99.4	9880.36	26.75
252	152.6	99.4	9880.36	64.75
5375	5375.0	0.0		$X^2 = 101.35$

statistic is found to be 101.35. For a 2×2 table the degrees of freedom from (1.14) is unity. At the 5% significance level the requisite value of the chi-square distribution from the table in Appendix A is 3.84. Clearly the value of the test statistic is such as to shed grave doubts on the truth of the null hypothesis of independence; since the present example involves a 2×2 table this may be interpreted as meaning that the proportion of males who die from tuberculosis of the respiratory system differs from the corresponding proportion for females. These proportions estimated from Table 1.1 are, males 0.929 and females 0.840.

It is well to note that the finding of a significant association by means of the chi-square test does not necessarily imply any causal relationship between the variables involved, although it does indicate that the reason for this association is worth investigating.

1.10 Summary

In this chapter the type of data with which this book is primarily concerned has been described. Testing for the independence of two qualitative variables by means of a chi-square test has been introduced. The chi-square test is central to this text, and in subsequent chapters many examples of its use in investigations in psychology, psychiatry and social medicine will be given. These sciences are as yet in the early stages of development, and studies in them are still characterized by a search for the variables basic to them. This search is often one for general relationships and associations – however amorphous they may appear at first – between the phenomena being studied, and here the chi-square test is often helpful.

2 × 2 contingency tables

2.1 Introduction

The simplest form of contingency table, namely that arising from two dichotomous variables and known as a 2 × 2 table, was introduced in the preceding chapter. In this chapter such tables are considered in greater detail.

Data in the form of 2 × 2 tables occur very frequently in social science, educational and medical research. Such data may arise in several ways. For example, they can occur when a total of, say, N subjects is sampled from a population and each individual is classified according to two dichotomous variables. Table 2.1 illustrates such a sampling scheme, 100 males being classified with respect to age, say above or below 40, and with respect amount of smoking, say above and below 20 cigarettes a day.

Such data may also arise in an investigation where a **predetermined** number of individuals in each of the categories of one of the variables are sampled, and for each sample the number of individuals in each of the categories of the second variable assessed. For example, in an investigation into the frequency of the side-effect nausea with a particular drug, 50 subjects were given the drug and 50 subjects given a placebo, and the number of subjects suffering from nausea in each sample recorded. Table 2.2 shows the outcome.

The analysis of such tables is by means of the chi-square test described in Chapter 1, a simplified form of which is available for 2 × 2 tables as described in the following section.

2.2 Chi-square test for a 2 × 2 table

The general 2 × 2 table may be written in the form shown in Table 2.3. The usual expression for computing the chi-square

Table 2.1   *Smoking and age example*

		Age Under 40	Age Over 40	
	Less than 20/day	50	15	65
Amount of smoking				
	More than 20/day	10	25	35
		60	40	100

Table 2.2   *Side-effects and drug example*

		Side-effect (nausea) Present	Side-effect (nausea) Absent	
	Drug given	15	35	50
Treatment				
	Placebo given	4	46	50
		19	81	100

Table 2.3   *General 2 × 2 contingency table*

		Variable A Category 1	Variable A Category 2	
	Category 1	a	b	$a + b$
Variable B				
	Category 2	c	d	$c + d$
		$a + c$	$b + d$	$N = a + b + c + d$

statistic, that is:

$$\sum \frac{(\text{observed frequency} - \text{expected frequency})^2}{\text{expected frequency}} \tag{2.1}$$

reduces the general 2 × 2 table to the following simplified form:

$$X^2 = \frac{N(ad - bc)^2}{(a + b)(c + d)(a + c)(b + d)}. \tag{2.2}$$

As was seen in Chapter 1, the significance of this statistic is judged by referring it to the tabulated chi-square values with one degree of freedom. For example, applying (2.2) to the data of Table 2.2 gives:

$$X^2 = \frac{100 \times (15 \times 46 - 35 \times 4)^2}{50 \times 50 \times 19 \times 81}$$

$$= 7.86.$$

At the 5% level the tabulated χ^2 value for 1 d.f. is 3.84. The calculated value is greater than this (its associated p value is 0.005), so some doubt is shed on the hypothesis that the occurrence of side-effects is independent of the treatments (drug and placebo) involved. The result may be interpreted as indicating that the proportion of people who suffer from nausea when treated with the drug is different from the proportion of people who suffer from nausea when given a placebo. In fact, the proportion of people suffering from nausea when treated with the drug ($\hat{p}_1 = 0.30$) appears to be considerably higher than the proportion of placebo-treated patients experiencing the side-effect ($\hat{p}_2 = 0.08$). An estimate of the standard error of the observed difference, $\hat{p}_1 - \hat{p}_2$, of the two proportions is given by

$$\left(\frac{\hat{p}_1(1 - \hat{p}_1)}{n_1} + \frac{\hat{p}_2(1 - \hat{p}_2)}{n_2} \right)^{1/2}. \tag{2.3}$$

This expression enables an approximate **confidence interval** (see Mood and Graybill, 1963, Ch. 11) to be found for the difference in the two proportions; in the nausea example the standard error given by (2.3) takes the value 0.0753 leading to an approximate 95% interval (0.0684, 0.3716).

2.3 Yates's continuity correction

In deriving the distribution of the statistic X^2, a **continuous** probability distribution, namely the chi-square distribution, is being used as an approximation to the **discrete** probability of observed frequencies, namely the multinomial distribution. To improve this approximation, Yates (1934) suggested a correction which involves subtracting 0.5 from the positive discrepancies, (observed − expected), and adding 0.5 to the negative discrepancies, before these values are squared in the calculation of the chi-square test statistic. The correction may be incorporated directly into (2.2), which then becomes:

$$X^2 = \frac{N(|ad - bc| - 0.5N)^2}{(a + b)(c + d)(a + c)(b + d)}. \qquad (2.4)$$

This is known as the chi-square value corrected for continuity. In (2.4) the term $|ad - bc|$ means 'the absolute value of $(ad - bc)$', that is the numerical value of the expression irrespective of its sign.

Conover (1968, 1974) has questioned the routine use of Yates's correction, but Mantel and Greenhouse (1968), Fleiss (1973) and Mantel (1974) reject Conover's arguments. In general the evidence for applying the correction seems convincing and hence its use is recommended. If the sample size is reasonably large the correction will have little effect on the value of the chi-square statistic.

2.4   Small expected frequencies: Fisher's exact test for 2 × 2 tables

As mentioned briefly in Chapter 1, one of the assumptions made when deriving the chi-square distribution as an approximation to the distribution of the test statistic in (1.11), is that the expected frequencies should not be too small since otherwise the approximation may not be good. The problem will be discussed in the next chapter, but in the case of 2 × 2 tables with 'small' expected frequencies, say five or less, then **Fisher's exact test** may be used as an alternative to the chi-square test.

2.4.1   Fisher's test for 2 × 2 tables

Fisher's exact test for a 2 × 2 contingency table does not use the chi-square approximation at all. Instead the exact probability distribution of the observed frequencies is employed. For fixed marginal totals the required distribution is easily shown to be that associated with sampling without replacement from a finite population, namely a **hypergeometric distribution** (see Mood and Graybill, 1963, Ch. 3). Assuming that the two variables are independent, the probability (P) of obtaining any particular arrangement of the frequencies a, b, c and d (Table 2.3), when the marginal totals are as given, is:

$$P = \frac{(a + b)!(a + c)!(c + d)!(b + d)!}{a!b!c!d!N!}, \qquad (2.5)$$

where $a!$ – read 'a factorial' – is the shorthand method of writing the product of a and all the whole numbers less than it, down to unity; for example:

$$5! = 5 \times 4 \times 3 \times 2 \times 1 = 120.$$

(By definition the value of 0! is unity.) Fisher's test now employs (2.5) to find the probability of the arrangement of frequencies actually obtained, *and* that of every other arrangement giving as much or more evidence of association, always keeping in mind that the marginal totals are to be regarded as fixed. The sum of these probabilities is then compared with the chosen significance level, α; if it is greater than α there is insufficient evidence for any association between the variables; if it is less than α then the hypothesis of independence is thrown into doubt. A numerical example will help to clarify the procedure.

2.4.2   *Numerical example of Fisher's exact test for 2 × 2 tables*

In a broad general sense psychiatric patients can be classified as psychotics or neurotics. A psychiatrist whilst studying the symptoms of a random sample of 20 patients from each of these diagnostic groups found that whereas six patients amongst the neurotics had suicidal feelings, only two psychotics suffered in this way; using these data, the psychiatrist wished to assess whether or not there was evidence of an association between diagnosis and suicidal feelings. The data are shown in Table 2.4. The hypothesis in this case is that the presence or absence of suicidal feelings is independent of the type of patient involved or equivalently that the proportion of psychotics with suicidal feelings is equal to the proportion of neurotics with the symptom.

The expected frequencies under the hypothesis of independence are shown in parentheses in Table 2.4; two of these expected values are below 5 and consequently the hypothesis will be tested by means of Fisher's exact test rather than by the chi-square statistic. Using (2.5), the probability of the observed table is calculated; it is

$$P_2 = \frac{8! \times 32! \times 20! \times 20!}{2! \times 6! \times 18! \times 14! \times 20!} = 0.095\,760.$$

(The subscript to P refers to the smallest of the frequencies, a, b, c, d, in this case the value 2.)

Returning to Table 2.4 and keeping in mind that the marginal frequencies are to be taken as fixed, the frequencies in the body of

Table 2.4 *The incidence of 'suicidal feelings' in samples of psychotic and neurotic patients*

	Type of patient		
	Psychotics	Neurotics	
Suicidal feelings	2(4)	6(4)	8
No suicidal feelings	18(16)	14(16)	32
	20	20	40

Table 2.5 *More exteme cell frequencies than those observed.*

(a)			(b)		
1	7	8	0	8	8
19	13	32	20	12	32
20	20	40	20	20	40

the table can be rearranged in two ways. Both would represent, if they had been observed, more extreme discrepancies between the groups with respect to the symptom. These arrangements are shown in Table 2.5.

Substituting in turn the values in Table 2.5(a) and 2.5(b) in (2.5) leads to:

$$P_1 = 0.020\,160, \text{ for Table 2.5(a),}$$
$$P_0 = 0.001\,638, \text{ for Table 2.5(b).}$$

Therefore the probability of obtaining the observed result, that is Table 2.4, or one more suggestive of a departure from independence is given by:

$$P = P_2 + P_1 + P_0$$
$$= 0.095\,760 + 0.020\,160 + 0.001\,638$$
$$= 0.117\,558.$$

This is the probability of observing amongst the eight patients suffering from suicidal feelings, that two or fewer are psychotics when the hypothesis of the equality of the proportions of psychotics and neurotics having the symptom, in the populations from which the samples were taken, is true; its value indicates that a discrepancy between the group as large or larger than that obtained might be expected to occur by chance about one in ten times even when there was no association between diagnosis and suicidal feelings. Since the probability values is larger than the commonly used significance levels (0.05 or 0.01), the data give no evidence that psychotics and neurotics differ with respect to the symptom. Indeed in this case, since P_2 is itself greater than 0.05, the computation could have ended before evaluating P_1 and P_0.

A significant result from Fisher's test indicates departure from the null hypothesis in a **specific** direction, in contrast to the chi-square test which assesses departures from the hypothesis in either direction. In the psychiatric groups example, the former is used to decide whether the proportions of patients in the two groups having suicidal feelings are equal or whether the proportion of psychotics with the symptom is **less** than the proportion of neurotics. The usual chi-square statistic, however, tests whether the proportions are **equal** or **unequal** without regard to the direction of the inequality. In other words, Fisher's test is **one-tailed** whereas the chi-square test is **two-tailed**. (For a detailed discussion of one-tailed and two-tailed tests, see Mood and Graybill, 1963, Ch. 12.) In the case where the sample sizes in each group are the same (as they are in the example above), the probability obtained from Fisher's test may be doubled to give the equivalent of a two-tailed test. This gives $P = 0.235\,12$. It is of interest to compare this value with the probability that would be obtained using the chi-square test. First we calculate the chi-square statistic for Table 2.4, using (2.2) to give:

$$X^2 = \frac{40(2 \times 14 - 18 \times 6)^2}{20 \times 20 \times 8 \times 32} = 2.50.$$

The exact value of the probability of obtaining a value of chi-square with one d.f. as large or larger than 2.50 is 0.113 85. Now we calculate the chi-square statistic with Yates's continuity correction applied using (2.4):

$$X^2 = \frac{40(|28 - 108| - 20)^2}{20 \times 20 \times 81 \times 32} = 1.41.$$

In this case the corresponding probability is 0.235 72. Since the comparable probability obtained from Fisher's test is 0.235 12 the efficacy of Yates's correction is clearly demonstrated.

The probability values found above from the chi-square test, the

Yates's corrected statistic and Fisher's test reflect a number of general points about the three tests when applied to small or moderate sized samples:

(1) Yates's corrected chi-square gives a probability value similar to that obtained from Fisher's exact test.
(2) Yates's corrected chi-square and Fisher's exact test give p values that are more conservative (i.e. larger) than those obtained from the chi-square test.

(In large samples it is well known that all three methods are equivalent.)

Because it is apparently conservative, Fisher's exact test has often been the subject of severe criticism. (A recent example is the paper by D'Agostino *et al.*, 1988.) Much of the criticism centres around the **conditional** nature of Fisher's test, assuming as it does that both sets of marginal totals are fixed. The arguments are of a subtle and technical nature but primarily involve whether or not the marginal totals in a table (however generated), provide any information on the existence of an association. Yates (1984) argues strongly that they do not and consequently concludes that Fisher's exact test **is** appropriate for the analysis of 2 × 2 tables. Cox (1984), Barnard (1984) and Little (1989) provide further support for this view.

2.4.3   *The power of Fisher's exact test for 2 × 2 tables*

The power of a statistical test (see, for example, Mood and Graybill, 1963, Ch. 12) is equal to the probability of rejecting the null hypothesis when it is untrue or, in other words, the probability of making a correct decision when applying the test. Obviously the power of any test used should be as high as possible. Several workers have investigated the power of Fisher's test and have shown that large sample sizes are needed to detect even moderately large differences between the two proportions. For example, Bennett and Hsu (1960) give a number of tables which allow the power of Fisher's test to be calculated in particular cases; their results show that for the data in Table 2.4 where sample sizes of 20 from each group (that is 20 neurotics and 20 psychotics) are involved, a test at the 5% level has a power of only 0.53 when the population values of the proportions of people with suicidal feelings in each group are 0.5 and 0.2. Therefore in almost half the cases of performing the test with these sample sizes the conclusion will be that there is **no** difference between the incidence of the symptom amongst psychotic and neurotic patients. The results of Gail and Gart (1973) show further that in this particular example a sample of 42 patients in each group would be needed to reach a power of 0.9 of detecting a difference in the population proportions. To detect small differences between the proportions, the latter authors show that relatively large sample sizes may be needed. For example, if the values of the population proportions were 0.8 and 0.6, a sample of 88 individuals from each category would be needed to achieve a power of 0.9, that is to have a 90% chance of detecting the difference.

2.5 McNemar's test for correlated proportions in a 2 × 2 table

One-to-one matching is frequently used by research workers to increase the precision of a comparison. The matching is usually done on variables such as age, sex, weight, IQ, and the like, such information usually being readily available. Two samples matched in this way must be thought of as **correlated** rather than independent; consequently the usual chi-square test is not strictly applicable for assessing the difference between frequencies obtained from such samples.

This appropriate test for comparing frequencies in matched samples is one due to McNemar (1955). As an introduction to it, consider Table 2.6 in which the presence or absence of some characteristic or attribute A for matched samples I and II is shown. Since interest lies in any difference between the sample with respect to A, frequencies in the N–E and S–W cells of the table are of little interest, since b refers to matched pairs both of which possess the attribute, which c refers to pairs both of which do not possess the attribute. The comparison is thus confined to the frequencies a and d, the former representing the number of matched pairs that possess the attribute if they come from sample I and do not possess it if

Table 2.6   *Frequencies in matched samples*

| | | Sample I | |
		A absent	A present
Sample II	A present	a	b
	A absent	c	d

they are from sample II, while the latter represents pairs where the converse situation holds. Under the hypothesis that the two samples do not differ as regards the attribute, a and d would be expected to be equal, with an estimated expected value $(a + d)/2$. Substituting the observed and expected values for the two cells into the usual formula for the chi-squared statistic, namely (2.1), leads to:

$$X^2 = \frac{(a - d)^2}{a + d}. \qquad (2.6)$$

If a correction for continuity is applied, the corresponding expression is:

$$X^2 = \frac{(|a - d| - 1)^2}{a + d}. \qquad (2.7)$$

This is McNemar's formula for testing an association in a 2 × 2 table when the samples are matched; under the hypothesis of no difference between the matched samples with respect to the attribute A, the test statistic has a chi-square distribution with one d.f. Examples will serve to illustrate the use of the test in practice.

2.5.1 Numerical examples of McNemar's test

A psychiatrist wished to assess the effect of the symptom 'depersonalization' on the prognosis of depressed patients. For this purpose 23 endogenous depressed patients who were diagnosed as being 'depersonalized' were matched one-to-one for age, sex, duration of illness and certain personality variables, with 23 endogenous depressed patients who were diagnosed as not being 'depersonalized'. On discharge, after a course of ECT, patients were assessed as 'recovered' or 'not recovered', and the results for the 23

Table 2.7 *Recovery of 23 pairs of depressed patients*

		Depersonalized patients		
		Not recovered	Recovered	
Patients not depersonalized	Recovered	5	14	19
	Not recovered	2	2	4
		7	16	23

pairs of patients are shown in Table 2.7. The value of a is 5 and of d is 2, leading to a chi-square value calculated from equation (2.7) of 0.57. Clearly this is not significant and consequently 'depersonalization' appears not to be associated with prognosis where endogenous depressed patients are concerned.

Although one-to-one matching is the matching procedure most commonly used, several investigators have considered the possibility of matching each case with several independent controls in the hope of increasing statistical efficiency. Ury (1975), for example, shows that this type of k-to-one matching has an efficiency of $2k/(k + 1)$ when compared to a matched pairs design. Consequently there is clearly a diminishing return in increasing k. Miettinen (1969) gives a test statistic for such designs which, when $k = 1$, is equivalent to McNemar's test.

McNemar's test is also applicable to situations in which the same subjects are observed on two occasions. For example, suppose that two drugs (A and B) are used in the treatment of depression, and are to be compared in terms of the possible side-effects, nausea. The drugs are given to the patients on two different occasions and the incidence of nausea recorded. Again we are dealing with correlated rather than independent observations (since the same subject receives both A and B) and consequently a comparison of the drugs will involve McNemar's test. Table 2.8 shows a set of data collected during such an investigation. Amongst these subjects 75 never experienced nausea, 13 subjects had nausea with A but not with B, 3 had nausea with B but not with A and 9 had nausea with both drugs. A McNemar test of these data gives

$$X^2 = \frac{(|3 - 13| - 1)^2}{16}.$$

$$= 5.06$$

Table 2.8   *Number of subjects showing nausea with drugs A and B*

| | | Drug A | | |
		No nausea	Nausea	
Drug B	Nausea	3	9	12
	No nausea	75	13	88
		78	22	100

This has an associated probability value of 0.02 and it appears therefore that there is some evidence that the incidence of nausea is different for the two drugs.

2.6 Cross-over designs

In the analysis of the drug and nausea data of Table 2.8, described in the previous section, one aspect of the study has been ignored, namely the order of drug administration. Most investigations of this type would involve a **cross-over** design, in which one group of subjects receive the drugs in the order AB, and other group receive them in the order BA. In most cases an equal number of subjects will be placed in each group. A more detailed analysis of such data will involve not only the investigation of drug differences, but also occasion differences and order effects, since in some situations the order of drug administration might have an appreciable effect on a subject' response. Gart (1969) has derived suitable tests for occasion and treatment effects which can be illustrated using the data given in Table 2.8, expanded to include information about patient group (see Table 2.9).

Gart's test for treatment effect involves the application of Fisher's exact test to the data from the pairs of observations giving unlike responses arranged as shown in Table 2.10(a), and the test for occasion effect involves a similar procedure applied to these observations, now arranged as shown in Table 2.10(b).

The corresponding arrangements for the nausea data are given in Tables 2.11(a) and 2.11(b); applications of Fisher's test to each of these gives p values as follows:

$$\text{Treatment effect: } p = 0.02,$$
$$\text{Occasion effect: } p = 0.50.$$

Table 2.9 *Nausea and drug with order of drug administration information*

	(0,0)	(0,1)	(1,0)	(1,1)	
Group 1 (AB)	40	1	7	2	50
Group 2 (BA)	35	6	2	7	50
	75	7	9	9	100

(r_1, r_2): r_1 and r_2 take values 0 and 1 corresponding to no nausea or nausea with first or second drug received.

Table 2.10 *Data arranged so as to test for (a) a treatment effect; (b) an order effect*

	(a) Drug order			(b) Drug order		
	(A, B)	(B, A)		(A, B)	(B, A)	
Nausea with first drug	y_a	y'_b	$y_a + y'_b$	Nausea with A $\quad y_a$	y'_a	$y_a + y'_a$
Nausea with second drug	y_b	y'_a	$y_b + y'_a$	Nausea with B $\quad y_b$	y'_b	$y_b + y'_b$
	n	n'	$n + n'$	n	n'	$n + n'$

Table 2.11

	(a) Drug order			(b) Drug order		
	(A, B)	(B, A)		(A, B)	(B, A)	
Nausea with first drug	$7(y_a)$	$2(y'_b)$	9	Nausea with A $\quad$ 7	6	13
Nausea with second drug	$1(y_b)$	$6(y'_a)$	7	Nausea with B $\quad$ 1	2	3
	8	8	16	8	8	16

These results indicate that the incidence of nausea is higher for drug A than for drug B, but that the incidence of nausea on the first and second occasions does not differ.

A complication with this approach is that both the tests described assume that there is no treatment × occasion interaction (sometimes referred to in this type of design as **carry-over effect**). Consequently it becomes of some importance to assess whether such an effect is present prior to applying Gart's tests for treatment and occasion differences. Hills and Armitage (1979) proposed a test for the treatment × occasion interaction which again involves the application of Fisher's test, but now to the **non-preference** responses. For the drug and nausea data the appropriate observations are shown in Table 2.12. The result of applying Fisher's test is a p value of 0.15; consequently the assumption of no treatment and occasion

Table 2.12 *Data for testing for carry-over effect for nausea data*

	(0,0)	(1,1)	
Group 1 (AB)	40	2	42
Group 2 (BA)	35	7	42
	75	9	84

interaction appears appropriate, making the treatment × occasion effect tests, described previously, acceptable.

A full account of the analysis of 2 × 2 cross-over designs for binary data is given by Kenward and Jones (1987).

2.7 Combining information from several 2 × 2 tables

In many studies a number of 2 × 2 tables all bearing on the same question may be available; if so, it becomes of interest to consider how to combine the tables in some way to make an overall test of the association between the row and column variables. For example, in an investigation into the occurrence of lung cancer among smokers and non-smokers, data may be obtained from several different locations or areas, and for each area the data might be arranged in a 2 × 2 table. Again, in an investigation of the occurrence of a particular type of psychological problem in boys and girls, data may be obtained from each of several different age groups, or from each of several different schools. The question is, how may the information from separate tables be pooled?

One obvious method which springs to mind is to combine all the data into a single 2 × 2 table for which a chi-square statistic is computed in the usual way. This procedure is legitimate only if corresponding proportions in the various tables are alike. Consequently, if the proportions vary from table to table, or we suspect that they vary, this procedure should not be used, since the combined data will not accurately reflect the information contained in the original tables. For example, in the lung cancer and smoking example mentioned previously, where data are collected from several different areas, it may well be the case that the occurrence of lung cancer is more frequent in some areas than in others. Armitage and Berry (1987) give an extreme example of the tendency of this procedure to create significant results.

Another technique which is often used is to compute the usual chi-square value separately for each table, and then to add them; the resulting statistic may then be compared with the value of chi-square from tables with g degrees of freedom where g is the number of separate tables. (This is based on the fact that the sum of g chi-square variables each with one degree of freedom is itself distributed as chi-square with g degrees of freedom.) This is also a poor method since it takes no account of the direction of the differences between the proportions in the various tables, and consequently lacks power in detecting a difference that shows up consistently in the same direction in all or most of the individual tables. More suitable techniques for combining the information from several 2 × 2 tables are described in the next section.

2.7.1   The $\sqrt{(X^2)}$ method

If the sample sizes of the individual tables do not differ greatly (say by more than a ratio of 2 to 1), and the values taken by the proportions are between approximately 0.2 and 0.8, then a method based on the sum of the square roots of the X^2 statistics, taking account of the signs of the differences in the proportions, may be used. It is easy to show that under the hypothesis that the proportions are equal, the X value for any of the 2 × 2 tables is approximately normal distributed with mean zero and unit standard deviation; consequently the sum of these X values for the complete set of tables (say g in number), is approximately normally distributed with mean zero and standard deviation $\sqrt{g}$. So as a test statistic for the hypothesis of no difference in the proportions in all the tables, Z given by

$$Z = \sum_{i=1}^{g} X_i / \sqrt{g} \qquad (2.8)$$

may be used, where X_i is the value of the square root of the usual X^2 statistic for the ith table with appropriate sign attached. To illustrate the method let us consider the data shown in Table 2.13 in which the incidence of malignant and benign tumours in the left and right hemispheres in the cortex is given. The problem is to test whether there is an association between hemisphere and type of tumour. Data for three sites in each hemisphere were available, but an earlier investigation had shown that there was no reason to suspect that any relationship between hemisphere and type of tumour

Table 2.13 *Incidence of tumours in the two hemispheres for different sites in the cortex*

Site of tumour	Benignant tumours	Malignant tumours	Proportion of malignant tumours	X^2	X
1. Left hemisphere	17	5	0.2273	1.7935	1.3392
Right hemisphere	6	5	0.4545		
	23	10			
2. Left hemisphere	12	3	0.2000	1.5010	1.2288
Right hemisphere	7	5	0.4167		
	19	8			
3. Left hemisphere	11	3	0.2143	2.003	1.4155
Right hemisphere	11	9	0.4500		
	22	12			

would differ from one site to another, so an overall assessment of the hemisphere–tumour relationship was indicated. For each of the three sites the number of patients (33, 27 and 34 respectively) is roughly equal, so we shall apply the $\sqrt{(X^2)}$ method to these data. The value of X^2 is first computed for each separate table. (Note that none of these is significant.) The square roots of these values are then obtained and the sign of the difference between the proportions is assigned to each value of X. For these data the difference between the proportions is in the same direction for each of the three tables, namely the proportion of malignant tumours in the right hemisphere is always higher than in the left. Consequently, the **same** sign is attached to each X value. (Whether this is positive or negative is, of course, immaterial.) The test statistic for the combined results is given by:

$$Z = \frac{1.339 + 1.229 + 1.415}{\sqrt{3}}$$

$$= 2.300.$$

This value is referred to the tables of the standard normal distribution and is found to be significant at the 5% level. Therefore, considering the three sites together suggests that there is an association between type of tumour and hemisphere.

If for these data the separate chi-square statistics, namely 1.793, 1.501 and 2.003, are added together, the value 5.297 is obtained. Testing this against a chi-square with three degrees of freedom gives a non-significant result. Clearly in this case where each of the differences is in the same direction, the $\sqrt{(X^2)}$ method is more powerful than those based on summing individual chi-square values.

2.7.2 Cochran's method

If the sample sizes and the proportions do not satisfy the conditions mentioned in the previous section, then addition of the X values tends to lose power. Tables that arise from very small sample sizes cannot be expected to be of as much use as those where the sample size is moderate to large in detecting a difference in the proportions, yet in the $\sqrt{(X^2)}$ method all tables receive the same weight. Where differences in the sample sizes are extreme, some method of **weighting** the results from individual tables is needed. Cochran (1954) suggested such a test, the test statistic Y being a weighted mean of the differences between the proportions in each table. Y is given by:

$$Y = \sum_{i=1}^{g} w_i d_i \Bigg/ \left(\sum_{i=1}^{g} w_i P_i Q_i \right)^{1/2}, \qquad (2.9)$$

where n_{i1} and n_{i2} are the sample sizes in the two groups for the ith

Table 2.14   *The incidence of tics in three samples of maladjusted children*

Age range		Tics	No tics	Total	Proportion with tics
5–9	Boys	13	57	70	0.1857
	Girls	3	23	26	0.1154
	Total	16	80	96	0.1667
10–12	Boys	26	56	82	0.3171
	Girls	11	29	40	0.2750
	Total	37	85	122	0.3033
13–15	Boys	15	56	71	0.2113
	Girls	2	27	29	0.0690
	Total	17	83	100	0.1700

table, p_{i1} and p_{i2} are the corresponding proportions and

$$P_i = (n_{i1}p_{i1} + n_{i2}p_{i2})/(n_{i1} + n_{i2}),$$
$$Q_i = (1 - P_i),$$
$$d_i = (p_{i1} - p_{i2}),$$
$$w_i = n_{i1} n_{i2}/(n_{i1} + n_{i2}).$$

Y is a weighted mean of the d_i values in which the weights used give greater importance to differences based on large than on small samples. Under the hypothesis that the population differences between proportions are zero for $i = 1, \ldots, g$, then the statistic Y is distributed normally with zero mean and unit variance.

To illustrate Cochran's procedure it will be applied to the data shown in Table 2.14 giving the incidence of tics in three age groups of boys and girls. In this case there are $g = 3, 2 \times 2$ tables and the various quantities needed to perform Cochran's test are as follows:

(a) Age range 5–9:

$$\begin{aligned}
n_{11} &= 70, & n_{12} &= 26, \\
p_{11} &= 0.1857, & p_{12} &= 0.1154, \\
P_1 &= 0.16671, & Q_1 &= 0.8333, \\
d_1 &= 0.0703, & w_1 &= 18.96.
\end{aligned}$$

(b) Age range 10–12:

$$\begin{aligned}
n_{21} &= 82, & n_{22} &= 40, \\
p_{21} &= 0.3171, & p_{22} &= 0.2750, \\
P_2 &= 0.3033, & Q_2 &= 0.6967, \\
d_2 &= 0.0421, & w_2 &= 26.89.
\end{aligned}$$

(c) Age range 13–15:

$$\begin{aligned}
n_{31} &= 71, & n_{32} &= 29, \\
p_{31} &= 0.2113, & p_{32} &= 0.0690, \\
P_3 &= 0.1700, & Q_3 &= 0.8300, \\
d_3 &= 0.1423, & w_2 &= 20.59.
\end{aligned}$$

Application of (2.9) gives

$$Y = \frac{18.96 \times 0.0703 + 26.89 \times 0.0421 + 20.59 \times 0.1423}{(18.96 \times 0.1667 \times 0.8333 + 26.89 \times 0.3033 \times 0.6967 + 20.59 \times 0.1700 \times 0.8300)^{1/2}}$$

$$= 1.61.$$

Referring this value to a normal curve it is found to correspond to a probability of 0.1074. Had the three age groups been combined and an overall chi-square test been performed, a value of 2.110 would have been obtained. This value corresponds to a probability of 0.2838 which is more than twice that given by Cochran's test, illustrating the greater sensitivity of the latter.

Another test procedure for examining a series of 2×2 tables is that suggested by Mantel and Haenszel (1959), but since it differs only marginally from Cochran's test, details will not be given here.

2.7.3 Further discussion of the $\sqrt{(X^2)}$ and Cochran's method of combining information from 2×2 tables

In cases where the relationship between the two variables is clearly very different in the separate 2×2 tables, neither the $\sqrt{(X^2)}$ nor Cochran's method is likely to be very informative. For example, suppose there were just two such tables with almost the same sample size; if the differences in the proportions of interest in the two tables were large, approximately equal in magnitude, but of **opposite** sign, then both the $\sqrt{(X^2)}$ and Cochran's statistic would be approximately zero and both would yield a non-significant result. Investigators need therefore to keep in mind that both procedures are essentially only useful for detecting departures from the null hypothesis due to differences in proportions which are **constant** from table to table. Application of either of the tests to sets of tables in which these differences vary greatly in magnitude and in direction should be avoided. In such cases combination of the tables in any way is not to be recommended and they are perhaps best dealt with by the methods to be described in Chapters 4 and 5.

An excellent extended account of the problem of combining evidence from fourfold tables is given in Fleiss (1973, Ch. 10).

2.8 Relative risks

So far in this chapter attention has been confined to significance tests for the hypothesis of no association in a 2×2 table. Important questions of **estimation** may, however, also arise, particularly when the null hypothesis is discarded. This is especially true of certain types of investigation; for example, in studying the aetiology of a disease it is often useful to measure the increased risk (if any) of incurring a specific disease if a certain factor is present. Suppose, for example, that in such studies the population of interest could be

Table 2.15

| | | Disease | | |
		Present $(+)$	Absent $(-)$	
Factor	Present $(+)$	P_1	P_3	$P_1 + P_3$
	Absent $(-)$	P_2	P_4	$P_2 + P_4$
		$P_1 + P_2$	$P_3 + P_4$	1

summarized in terms of the entries in Table 2.15. If the proportions in this table were known then the risk of having the disease for those individuals having the factor present would be

$$P_1/(P_1 + P_3) \qquad (2.10)$$

and for those individuals not having the factor present it would be

$$P_2/(P_2 + P_4). \qquad (2.11)$$

Estimation of the risks in (2.10) and (2.11) and associated confidence intervals is made from a sample of observations classified according to the presence or absence of both the disease and the factor – see Table 2.16. In cases where the data arise from a **cohort** or **prospective** study, in which individuals with and without the factor are followed for some period of time and the number of occurrences of the disease in each group noted, the estimates are found simply as $a/(a + b)$ and $c/(c + d)$.

An alternative to the cohort study is a **case-control** or **retrospective** study in which individuals known either to have the disease or not are followed backwards in time to decide whether or not the factor of interest has been present. For such a study estimation of the risks of concern is **not** possible. In this case what can be estimated is the

Table 2.16

| | | Disease | | |
		$+$	$-$	
Factor	$+$	a	b	$a + b$
	$-$	c	d	$c + d$
		$a + c$	$b + d$	N

risk of the factor being present amongst people with the disease $a/(a + c)$ and similarly for those not having the disease $b/(b + d)$.

In many situations involving this type of example the proportion of subjects having the disease will be small; consequently P_1 will be small compared to P_3, and P_2 will be small compared to P_4. The ratio of the risks given by (2.10) and (2.11) then very nearly becomes

$$\frac{P_1 P_4}{P_2 P_3}. \tag{2.12}$$

This is usually referred to as the **odds ratio**, and acts as an approximation to the relative risk; it is often also called simply the **relative risk** and denoted by ψ. If the corresponding **sample** frequencies are as shown in Table 2.16, then an estimate of ψ is given by

$$\hat{\psi} = \frac{ad}{bc}. \tag{2.13}$$

(This ratio can be estimated from both prospective and retrospective studies, although in the latter the $a + c$ individuals with the disease must be a random, unbiased sample of all cases of the disease, and the $b + d$ individuals not having the disease must be a similar sample of all people without the disease.)

In general, in addition to a **point estimate** of ψ as given by (2.13) a **confidence interval** would usually be required. (See Mood and Graybill, 1963, Ch. 11.) Such an interval is most easily found by initially considering $\ln \hat{\psi}$ since its variance may be estimated very simply as follows:

$$\text{vâr}(\ln \hat{\psi}) = \frac{1}{a} + \frac{1}{b} + \frac{1}{c} + \frac{1}{d}. \tag{2.14}$$

Consequently an approximate 95% confidence interval for $\ln \psi$ is given by

$$\ln \hat{\psi} \pm 1.96 \times \sqrt{[\text{vâr}(\ln \hat{\psi})]}. \tag{2.15}$$

The required confidence interval for ψ would now be found by taking exponentials of the two values given by (2.15). A numerical example will help to clarify this procedure.

2.8.1 Calculating a confidence interval for relative risk

Suppose that of the members of a given population equally exposed to a virus infection, a percentage (which can be assumed to contain

Table 2.17   *Incidence of virus infection*

	Not infected	Infected	
Not inoculated	130 (d)	20 (c)	150
Inoculated	97 (b)	3 (a)	100
	227	23	250

a fair cross-section of the population as a whole) has been inoculated. After the epidemic has passed, a random sample of people is drawn from the population and the numbers of inoculated and uninoculated that have escaped infection are recorded giving the data shown in Table 2.17.

It is clear from these data that the proportion of uninoculated people that was infected by the virus is considerably larger than the proportion of inoculated infected so that the risk of infection is less for those inoculated than those uninoculated. A confidence interval for relative risk can be used to quantify this difference.

From (2.13) $\hat{\psi}$ is found to be 0.201, so that $\ln \hat{\psi} = -1.60$. The estimated variance of $\ln \hat{\psi}$ is obtained from (2.14) as 0.401, giving an approximate 95% confidence interval for $\ln \hat{\psi}$ of

$$-1.60 \pm 1.96 \times 0.63,$$

that is

$$-2.83 \text{ to } -0.37.$$

Taking exponentials of these two limits gives the required 95% confidence interval for ψ of 0.06 to 0.69. Consequently the risk that an inoculated person will be infected by the virus is likely to be at most 69% of that of an uninoculated person and it may be as low as 6%.

Frequently an estimate of relative risk is made from each of a number of subsets of the data, and interest may be centred on combining these various estimates. One approach is to take separate estimates of $\ln \psi$ and weight them by the reciprocal of their variance (formula (2.14)). The estimates may then be combined by taking a weighted mean. An alternative pooled estimate of relative risk is given by Mantel and Haenszel (1959). This, using an obvious nomenclature for the various 2 × 2 tables involved, is given by

$$\hat{\psi}_{\text{pooled}} = \frac{\sum (a_i d_i / N_i)}{\sum (b_i c_i / N_i)}. \tag{2.16}$$

Table 2.18 *Number of cases of bronchitis by level of organic particulates in the air and by age (taken with permission from Somes and O'Brian, 1985)*

Age	Organic particulates level	Bronchitis Yes	No	Total
15–24	High	20	382	402
	Low	9	214	223
23–39	High	10	172	182
	Low	7	120	127
40 +	High	12	327	339
	Low	6	183	189

To illustrate the estimation of this pooled relative risk the data given in Table 2.18 (adapted with permission from Somes and O'Brian, 1985) will be used.

Mantel and Haenszel's estimate is given by

$$\hat{\psi}_{\text{pooled}} = \frac{6.85 + 3.88 + 4.16}{5.50 + 3.90 + 3.72}.$$

$$= 1.13$$

A measure of the variance of $\hat{\psi}_{\text{pooled}}$ is given by Robins *et al.* (1986) and this facilitates the construction of confidence intervals. This variance is given by

$$\text{var}(\hat{\psi}_{\text{pooled}}) = \frac{\sum P_i R_i}{2R_+^2} + \frac{\sum (P_i S_i + Q_i R_i)}{2R_+ S_+} + \frac{\sum Q_i S_i}{2S_+^2}, \qquad (2.17)$$

where

$$P_i = (a_i + d_i)/n_i,$$
$$R_i = a_i d_i/n_i,$$
$$R_+ = \sum R_i,$$
$$Q_i = (b_i + c_i)/n_i,$$
$$S_i = b_i c_i/n_i,$$
$$S_+ = \sum S_i.$$

For the data in Table 2.18 the various terms needed to calculate this variance are as follows:

$$
\begin{array}{llll}
P_1 = 0.37 & R_1 = 6.85 & Q_1 = 0.63 & S_1 = 5.50, \\
P_2 = 0.42 & R_2 = 3.88 & Q_2 = 0.58 & S_2 = 3.90, \\
P_3 = 0.37 & R_3 = 4.16 & Q_3 = 0.63 & S_3 = 3.72,
\end{array}
$$

$$
R_+ = 14.89 \quad S_+ = 13.12,
$$

leading to

$$
\mathrm{var}(\hat{\psi}_{\mathrm{pooled}}) = \frac{5.70}{443.42} + \frac{(5.05 + 9.19)}{390.17} + \frac{8.07}{344.27}.
$$

$$
= 0.0727
$$

The relative risk is often estimated from a matched-pairs, case-control study; in such cases testing hypotheses about ψ is based on the binomial distribution **conditional** on the number of discordant pairs. The McNemar statistic is used as an approximation to the binomial test. Sample sizes and power of such studies are considered in Connett *et al.* (1987).

2.9 Guarding against biased comparisons

In Chapter 1 the need to use random or representative samples as a safeguard against obtaining biased results in an investigation was stressed. Now that a few examples are available to which to refer, some further discussion of the matter will be helpful.

An important advance in the development of statistical science was achieved when the advantages of design in experimentation were realized (Fisher, 1950). These advantages result from conducting an investigation in such a way that environmental effects and other possible factors, which might make interpretation of the results ambiguous, are kept under control. But in many investigations in social medicine and in survey work in general, where the data are often of a qualitative kind (and chi-square tests are commonly required), planned experiments are difficult to arrange (Taylor and Knowelden, 1957, Ch. 4). One of the problems is that the occurrence of the phenomenon being studied may be infrequent, so the time available for the investigation permits a retrospective study only to be undertaken. With such studies it is generally difficult to get suitable control data, and serious objections often arise to the samples one might draw, because of limitations in the population being sampled. Berkson (1946) has drawn attention to this point where hospital populations are concerned, and he has demonstrated that the subtle differential selection factors which operate in the referral of people to hospital are likely to bias the results of investigations based on samples from these populations. His main point can be illustrated best by an example.

Suppose an investigator wished to compare the incidence of tuberculosis of the lung in postmen and bus drivers. He or she might proceed by drawing two samples from the entrants to these occupations in a given month or year and do a follow-up study, with regular X-ray examinations, over a period of years to obtain the information required. He or she would, of course, be aware of the possibility that people, by reason of their family histories or suspected predispositions to special ailments, might tend to choose one occupation rather than the other, and might take steps to control for such possibilities and to eliminate other possible sources of bias. But suppose that, since time and the facilities at his or her disposal did not permit a prospective study to be carried out, he or she decided to obtain samples by consulting the files of a large hospital and extracting for comparison all the postmen and bus drivers found there. The data obtained might not give a true picture. For instance, it might be the case that bus drivers, by virtue of the special responsibility attached to their jobs, were more likely than postmen to be referred to hospital should tuberculosis be suspected. If this were so, a biased comparison would clearly result.

A biased comparison would also result were it the case that bus drivers, say, were prone to be affected by multiple ailments such as bronchitis and tuberculosis, or carcinoma of the lung and bronchitis, or all three, since these ailments would be likely to aggravate each other so that a bus driver might be referred to hospital for bronchial treatment and then be found to have tuberculosis. Were this a common occurrence then a comparison of postmen and bus drivers as regards the incidence of tuberculosis, based on such hospital samples, would not give true reflection of the incidence of the disease in these occupations in the community.

The relevance of the above discussion to the interpretation of results from investigations such as those reported in this chapter can now be examined. For instance, if we return to Table 2.4 it is clear that the chi-square test applied to the data in it yields an unbiased result only in so far as we can be sure that the hospital populations of psychotics and neurotics from which the samples are drawn are not affected by differential selection. In particular we would want to satisfy ourselves that 'suicidal feelings' did not play a primary part in the referral of neurotics to hospital in the first place. If it did the results gven by the chi-square test would be biased.

But it is well to add that Berkson (1946) notes certain conditions under which unbiased comparisons can be made between samples drawn from sources in which selective factors are known to operate. For instance, if the samples of postmen and bus drivers drawn from the hospital files are selected according to some other disease or characteristic unrelated to tuberculosis, say those who on entry to hospital were found to require dental treatment, then a comparison of the incidence of tuberculosis in these people would yield an unbiased result.

As a means of avoiding a biased comparison between samples from a biased source it might too be thought that a one-to-one matching of subjects from the populations to be compared would overcome difficulty. But clearly this could not act as a safeguard. In the example discussed earlier in the chapter, in which the effect of the symptom 'depersonalization' on the prognosis of endogenous depressed patients was assessed, were it the case – which is unlikely – that 'depersonalization' itself was a primary factor in causing depressed patients to come to hospital, then the result of the comparison made in that investigation would be open to doubt.

2.10 Summary

In this chapter the analysis of 2×2 contingency tables has been considered in some detail. However, some warning should be given against rushing to compute a chi-square for every 2×2 table which readers may meet. They should first have some grounds for thinking that the hypothesis of independence is of interest before proceeding to test it. Secondly, calculation of the chi-square statistics if often a time-filler and a ritual which may prevent them from thinking of the sort of analysis most needed. For example, in many cases of 2×2 tables arising from survey data, the need is for a measure of the degree of association rather than a statistical test for association per se. Such measures are discussed in the following chapter. In other cases, estimation of the relative risk may be what is required.

A detailed mathematical account of some other approaches to the analysis of 2×2 tables is available in Cox and Snell (1989).

$r \times c$ contingency tables

3.1 Introduction

The analysis of $r \times c$ contingency tables when either r or c or both are greater than 2, presents special problems not met in the preceding chapter. For example, the interpretation of the outcome of a chi-square test for independence in the case of a 2×2 table can clearly be made in terms of the equality or otherwise of two proportions. In the case of larger contingency tables having more than a single degree of freedom, such interpretation is not so straightforward and more detailed analyses may be necessary to decide just where in the table any departures from independence arise. Appropriate methods are discussed in this chapter. Firstly, however, a numerical example of the usual chi-square test of independence applied to a 3×3 table is given.

3.2 Numerical example of a chi-square test

Table 3.1 shows a set of data in which 141 individuals with brain tumours have been doubly classified with respect to type and site of tumour. The three types were A, benignant tumours; B, malignant tumours; C, other cerebral tumours. The sites concerned were I, frontal lobes, II, temporal lobes; III, other cerebral areas.

In this example $r = c = 3$, and the null hypothesis H_0 is that the site and type of tumour are independent. Under this hypothesis, the estimated expected frequencies may be calculated from (1.9). For example,

$$E_{11} = \frac{38 \times 78}{141} = 21.02,$$

similarly

$$E_{12} = \frac{38 \times 37}{141} = 9.97.$$

The complete set of expected values under independence is shown in Table 3.2.

Although frequencies such as 21.02 are obviously not possible, the terms after the decimal point are retained to increase the accuracy when calculating the chi-square test statistic. (Note that the marginal totals of the expected values are equal to the corresponding marginal totals of observed values, so that $E_{i.} = n_{i.}$ for $i = 1, \dots, r$, and $E_{.j} = n_{.j}$ for $j = 1, \dots, c$. That this must **always** be true is easily seen by summing equation (1.9) over either i or j.) Using (1.13) to calculate the test statistic for assessing independence gives

$$X^2 = \frac{(23.0 - 21.02)^2}{21.02} + \frac{(9.0 - 9.97)^2}{9.97} + \cdots + \frac{(17.0 - 13.83)^2}{13.83}$$

$$= 0.19 + 1.96 + \cdots + 0.72$$

$$= 7.84.$$

Table 3.1 has four degrees of freedom so the chi-square test statistic has an associated p value of 0.098. The data produce little evidence against the hypothesis that the two classifications are independent;

Table 3.1 *Incidence of cerebral tumours*

			Type		
		A	B	C	
Site	I	23	9	6	38
	II	21	4	3	28
	III	34	24	17	75
		78	37	26	141

Table 3.2 *Expected frequencies for the data of Table 3.1*

			Type		
		A	B	C	
Site	I	21.02	9.97	7.01	38
	II	15.49	7.35	5.16	28
	III	41.49	19.68	13.83	75
		78	37	26	141

consequently no association between site and type of tumour can be claimed (but see Davies, 1991).

3.3 Small expected frequencies

The derivation of the chi-square distribution as an approximation for the distribution of the test statistic, X^2, when the hypothesis of independence is true, is made under the assumption that the expected values are not too small. Typically this vague term has been interpreted as meaning that a satisfactory approximation is achieved when expected frequencies are five or more. This restriction appears, however, to be an arbitrary one based more upon tradition than either mathematical or empirical evidence, and there appears to be no more justification for the five-or-more rule than for, say, a one-or-more rule.

Cochran (1954) was the first to point out that the usual 'rule' is too stringent and suggested that if relatively few expectations are less than five (say one cell in five), a minimum expectation of unity is allowable. Even this suggestion may be too restrictive since work by Lewontin and Felsenstein (1965), Slakter (1966), Roscoe and Byars (1971) and Larntz (1978) shows that many of the expected values may be as unity without affecting the test greatly. Lewontin and Felsenstein give the following conservative rule for tables in which $r = 2$: 'The $2 \times c$ table can be tested by the conventional chi-square criterion if all the expectations are 1 or greater.' The authors point out that even this rule is extremely conservative and in the majority of cases the chi-square test may be used for tables with expectations in excess of 0.5 in the smallest cell.

Nevertheless for small, sparse or skewed data the asymptotic theory may not be valid, although it is often difficult to predict *a priori* whether a given data set may cause problems. An **exact** test of the hypothesis that the row and column classifications are independent in an $r \times c$ contingency table can be executed in principle by generalizing Fisher's exact treatment of the 2×2 table. The computational effort required, however, has, until recently, severely limited this approach. But within the last ten years, the advent of fast algorithms and the availability of inexpensive computing power has considerably extended the bounds where the exact test is feasible. Details of these algorithms are unfortunately outside the level of this text and interested readers are referred to Mehta and Patel (1983, 1986a, 1986b) for a full exposition. It is, however, possible to illustrate the differences that can occur between the exact approach and the usual procedure involving the chi-square distribution. For example consider the following 3×9 contingency table:

0	7	0	0	0	0	0	1	1
1	1	1	1	1	1	1	0	0
0	8	0	0	0	0	0	0	0

The chi-square statistic for this table takes the value 22.9. The corresponding p value is 0.1342. The true p value, however, is 0.0013. The exact analysis indicates that the row and column classifications are **not** independent. The asymptotic analysis fails to show this relationship.

A procedure that has been used almost routinely for many years to overcome the problem of small expected frequencies is the pooling of categories. Such a procedure may be criticized for several reasons. Firstly, a considerable amount of information may be lost by the combination of categories and this may detract greatly from the interest and usefulness of the study. Secondly, the randomness of the sample may be affected. The whole rationale of the chi-square test rests on the randomness of the sample and on the categories into which the observations may fall being chosen in advance. Pooling categories after the data are seen may affect the random nature of the sample with unknown consequences. Lastly the manner in which the categories are pooled can have an important effect on the inferences drawn. As an example, consider the following data set given by Baglivo *et al.* (1988):

	Column				
	1	2	3	4	5
Row 1	2	3	4	8	9
Row 2	0	0	11	10	11

When this table is tested for independence using the usual approximate methods, the significance level calculated is 0.086 which agrees with Fisher's exact probability to two significant digits, although a standard statistical package will issue a warning like 'Some of the expected values are less than 2, the test may not be appropriate.' If the first two columns are ignored however, the significance becomes 0.48, and if the first two columns are collapsed into the third, it is 1.00.

The practice of combining classification categories should be avoided and is no longer needed because of the availability of the exact tests referred to above.

3.4 Isolating sources of association in $r \times c$ tables

A significant overall chi-square test for an $r \times c$ contingency table indicates that the variables forming the table are not independent, but provides no information as to whether the lack of independence occurs throughout the table or only in a specific section. Various methods have been suggested for isolating the parts of a contingency table responsible for any departure from independence and a number of these are discussed in this section. (The situation may be thought of as analogous to that arising when using analysis-of-variance techniques where, having found that a set of means differ, it is required to identify in more detail which **particular** means differ.)

3.4.1 Partitioning $r \times c$ tables

Lancaster (1949) and Irwin (1949) have shown that the overall chi-square statistic for a contingency table can always be partitioned into as many components as the table has degrees of freedom. Each component chi-square value corresponds to a particular 2×2 table arising from the original, and each component is independent of the others. The basic idea is to identify subtables that break up the chi-square statistic into more interpretable pieces, enabling those categories responsible for a significant overall chi-square value to be identified. The method will be illustrated using the data on retarded activity among psychiatric patients given in Table 3.3.

The usual chi-square test for independence applied to these data gives a value of 5.70 which with two degrees of freedom just falls short of the 5% level of significance ($p = 0.058$). An examination of the data, however, suggests that although the incidence of the symptom for the first two groups is very alike, it occurs more frequently amongst these groups than in the neurotic group. It is tempting to combine the affective disorders and the schizophrenics

Table 3.3   *Retarded activity amongst psychiatric patients*

	Affective disorders	Schizo- phrenics	Neurotics	
Retarded activity	12	13	5	30
No retarded activity	18	17	25	60
	30	30	30	90

and perform a chi-square test on the resulting 2×2 table. Indeed if this is done a value of $X^2 = 5.62$ is obtained, which has an associated p value of 0.02. But such a procedure, carried out purely in the hope of achieving a significant result after the test on the original data failed to reject the hypothesis of independence, would be quite unjustified and contrary to good statistical practice. In Table 3.3 the expected frequencies are all ten or more so there is no justification for combining groups under this pretext.

Methods of partitioning the overall chi-square value provide a means of examining contingency table data in greater detail and of obtaining more sensitive tests of association than could have been obtained otherwise. Kimball (1954) has supplied convenient formulae for obtaining the chi-square values corresponding to the partitioning method proposed by Lancaster and Irwin. To introduce these formulae consider a contingency table with r rows and c columns having the general form shown in Table 3.4. (This nomenclature is used here so that Kimball's formula can be illustrated more clearly.)

The value of X^2 calculated in the usual way from this table has $(r-1)(c-1)$ degrees of freedom and the first step in the partitioning process is to construct the $(r-1)(c-1)$ fourfold tables from which the components of chi-square are calculated. In the case of a 2×3 table, for example, the two fourfold tables may be constructed as

Table 3.4

$a_1 a_2$	.	.	.	a_c	A
$b_1 b_2$	.	.	.	b_c	B
.					.
.				.	
$n_1 n_2$	.	.	.	n_c	N

follows:

$$\frac{a_1 \mid a_2}{b_1 \mid b_2}$$

$$\frac{(a_1 + a_2) \mid a_3}{(b_1 + b_2) \mid b_3}$$

For the data in Table 3.3, the corresponding fourfold tables are

$$\frac{12 \mid 13}{18 \mid 17}$$

$$\frac{25 \mid 5}{35 \mid 25}$$

The decision as to **which** particular pair of columns to combine has to be made by the investigator in the light of prior knowledge about the classification categories concerned. (The decision about which columns are to be combined should, of course, always be made **before** examining the data to be analysed.)

Kimball's formulae for partitioning the overall chi-square value in the case of a $2 \times c$ table are obtained by giving t the values $1, 2, \ldots, (c-1)$ in turn in the general formula:

$$X_t^2 = \frac{N^2[b_{t+1}S_t^{(a)} - a_{t+1}S_t^{(b)}]^2}{ABn_{t+1}S_t^{(n)}S_{t+1}^{(n)}}, \tag{3.1}$$

where

$$S_t^{(a)} = \sum_{i=1}^{t} a_i, \quad S_t^{(b)} = \sum_{i=1}^{t} b_i, \quad S_t^{(n)} = \sum_{i=1}^{t} n_i,$$

and the other symbols are as defined in Table 3.4. Each X_t^2 value is a one degree of freedom component of the overall X^2 so that

$$X^2 = X_1^2 + X_2^2 + \cdots + X_{(c-1)}^2. \tag{3.2}$$

For the data in Table 3.3, $c = 3$ and the overall chi-square statistic will be partitioned into two components X_1^2 and X_2^2 each having a single degree of freedom. Substituting the values $t = 1$ and $t = 2$ into (3.1) gives the following simplified formulae for the two-component chi-square statistics:

$$X_1^2 = \frac{N^2(a_1b_2 - a_2b_1)^2}{ABn_1n_2(n_1 + n_2)}, \qquad (3.3)$$

$$X_2^2 = \frac{N^2[b_3(a_1 + a_2) - a_3(b_1 + b_2)]^2}{ABn_3(n_1 + n_2)(n_1 + n_2 + n_3)}. \qquad (3.4)$$

Substituting the values in Table 3.3 in these formulae gives $X_1^2 = 0.075$ and $X_2^2 = 5.625$. Each component has a single degree of freedom; the first is not significant, but the second is significant beyond the 2.5% level. Partitioning of the overall chi-square value, which itself was not significant, has led to a more sensitive test which allows the conclusion that whereas the first two groups of patients do not differ where the symptom 'retarded activity' is concerned, the two groups combined differ significantly from the third.

It should be noted that (3.1) differs slightly from the usual form of the chi-square test statistic for a 2×2 table (see Chapter 2), in so far as it has an additional term in the denominator, whilst instead of N in the numerator it has N^2. The formula as it stands has no correction for continuity but in cases where such a correction is considered appropriate it can be applied in the usual way. If this is done, additivity (i.e. (3.2)) is no longer exact, but the discrepancy between X^2 and the sum of its component parts is generally negligible.

The general formula for finding the components of the overall chi-square statistic in the case of a table with $r > 2$ is also given by Kimball but since it is very cumbersome the reader is referred to the original article.

3.4.2 *Partitioning $2 \times c$ tables into non-independent 2×2 tables*

The Lancaster and Irwin method described in the previous section subdivides the overall chi-square value into independent components such that formula (3.2) holds. Many researchers, however, wish to test specific hypotheses about particular sections of a contingency table. For example, in a trial of several drugs where the response variable of interest is dichotomous, of particular interest might be a comparison of each drug with a placebo. The chi-square values arising from each 2×2 table are not independent of each other, and it would be unsatisfactory to test them simply as chi-square variables with one degree of freedom, since such a procedure could have a serious effect on the value of the significance level in the sense of claiming more differences between placebo and drugs than the data actually merit. Brunden (1972, 1987) shows that a more reasonable approach is to adjust the significance level of each test from α to α' where

$$\alpha' = \frac{\alpha}{c - 1} \tag{3.5}$$

and c is the number of columns in the table. Essentially, each of the $c - 1$ comparisons performed is tested at a more stringent significance level.

To illustrate the procedure consider an example in which five drugs for treating depression are to be compared. Six samples of 30 depressed patients are taken and each patient is given one of the five drugs or a placebo; at the end of a two week period each patient is rated by an experienced clinician as being 'less depressed' or 'same or worse' than before receiving treatment. The results are shown in Table 3.5

The overall chi-square test for the data in Table 3.5 gives a value of 14.78, which is significant at the 5% level. Of particular interest here is which drugs differ from the placebo and exploring this question involves forming five separate 2×2 tables and calculating the chi-square test statistic for each (Tables 3.6 (a) to (e)).

Taking $\alpha = 0.05$, the required significance level at which to test each table, calculated from (3.5), is $\alpha = 0.05/5 = 0.01$. Only the data sets corresponding to drugs 2 and 5 reach this level of significance. [Note here that the component chi-square values do not sum to the overall chi-square value of 14.78.]

Rodger (1969) describes other methods whereby specific hypotheses of interest may be tested in $2 \times c$ tables.

Table 3.5   *Treatment of depression*

	Placebo	Drug 1	Drug 2	Drug 3	Drug 4	Drug 5	
Improved	8	12	21	15	14	19	89
Same or worse	22	18	9	15	16	11	91
	30	30	30	30	30	30	180

Table 3.6

(a)	Placebo	Drug 1		
Improved	8	12	20	$X^2 = 1.20$
Same or worse	22	18	40	
	30	30	60	

(b)	Placebo	Drug 2		
Improved	8	21	29	$X^2 = 11.28$
Same or worse	22	9	31	
	30	30	60	

(c)	Placebo	Drug 3		
Improved	8	15	23	$X^2 = 3.45$
Same or worse	22	15	37	
	30	30	60	

(d)	Placebo	Drug 4		
Improved	8	14	22	$X^2 = 2.58$
Same or worse	22	16	38	
	30	30	60	

(e)	Placebo	Drug 5		
Improved	8	19	27	$X^2 = 8.15$
Same or worse	22	11	33	
	30	30	60	

3.4.3   The analysis of residuals

A further procedure which is often helpful in identifying the cells of a contingency table responsible for a significant overall chi-square value is inspection of the deviations of observed from expected values measured in some appropriate way. The most obvious way of defining such a **residual** would be to take (observed value − expected value) for each cell. This would, however, be very unsatisfactory since a difference of fixed size is clearly more important for smaller samples. A more appropriate residual would be e_{ij} given by:

$$e_{ij} = (n_{ij} - E_{ij})/\sqrt{E_{ij}}, \qquad (3.6)$$

where E_{ij} is obtained from (1.9) as $n_i.n_{.j}/N$. These terms are usually known as **standardized residuals** and are such that the chi-square test statistic is given by

$$X^2 = \sum_{i=1}^{r} \sum_{i=1}^{c} e_{ij}^2. \qquad (3.7)$$

It is tempting to think that the size of these residuals may be judged by comparison with standard normal percentage points (for example, ± 1.96). Unfortunately it can be shown that the variance of e_{ij} is always less than or equal to one, and in some cases considerably less than one. Consequently the use of standardized residuals for detailed examination of a contingency table may often give conservative indications of cells having lack of fit.

At the cost of some extra calculation a more precise analysis may be achieved by using **adjusted residuals**, d_{ij}, as suggested by Haberman (1973). These are defined as follows:

$$d_{ij} = e_{ij}/\sqrt{[(1 - n_i./N)(1 - n_{.j}/N)]}. \qquad (3.8)$$

When the variables forming the contingency table are independent, the adjusted residuals are approximately normally distributed with mean zero and standard deviation 1.

Table 3.7 *Piston-ring failures in four compressors. Reproduced from Haberman (1973) by permission of the Biometrics Society*

		North	Leg Centre	South	
Comp. no.	1	17	17	12	46
	2	11	9	13	33
	3	11	8	19	38
	4	14	7	28	49
		53	41	72	166

Table 3.8   (a) *Adjusted residuals for piston-ring failures; (b) standardized residuals for piston-ring failures.
Reproduced from Haberman (1973) by permission of
the Biometrics Society*

		North	Leg Centre	South
		(a)		
	1	0.86	2.27	−2.78
Comp. no.	2	0.19	0.38	−0.52
	3	−0.45	−0.59	0.94
	4	−0.60	−2.01	2.32
		(b)		
	1	0.60	1.67	−1.78
Comp. no.	2	0.14	0.30	−0.35
	3	−0.32	−0.45	0.62
	4	−0.41	−1.47	1.46

To illustrate the use of adjusted residuals in data analysis, the
data shown in Table 3.7 (taken with permission from Haberman,
1973) will be used. The observations in this table represent the number
of piston ring failures in each leg of four compressors at a chemical
plant. The chi-square statistic for the table takes the value 11.7 with
six degrees of freedom showing some relatively weak evidence of an
association between compressor and location of piston ring failure
($p = 0.069$). The standardized and adjusted residuals are shown in
Table 3.8 (a) and (b). Four adjusted residuals have absolute values
greater than 2, the largest corresponding to the south leg of
compressor 1. In this example, use of adjusted rather than
standardized residuals has a considerable effect on the analysis. The
largest absolute standardized residual is only 1.78, so the size of
these residuals gives no evidence of departure from independence.

3.5   Examining contingency tables graphically: correspondence analysis

Correspondence analysis attempts to display graphically the
relationship between the variables forming a contingency table by
deriving a set of coordinates representing the row and column
categories of the table. The correspondence analysis coordinates are
analogous to those derived from a principal components analysis of
continuous, multivariate data (see Everitt and Dunn, 1991), except

that they are derived by partitioning the total chi-square statistic for the table, rather than total variance. In this section only a brief account of the method is given; a full account of correspondence analysis is available in Greenacre (1984).

The coordinates representing the row and column categories are derived essentially from the matrix, E, containing residuals from fitting the independence model; that is, a matrix with elements

$$e_{ij} = \frac{n_{ij} - E_{ij}}{\sqrt{E_{ij}}}. \qquad (3.9)$$

A process known as the **singular value decomposition** of E (see Everitt and Dunn, 1991), leads to two sets of coordinates, one set representing the rows of the tables, the other the columns. Generally the first, or more often, the first two, coordinates for each row category and for each column category are used to display the table graphically, since it is these which usually account for a large proportion of the total chi-square value. (In correspondence analysis X^2/N is often known as 'inertia', and the adequacy of a one- or two-dimensional representation of the residuals is judged by the proportion of the inertia explained by each dimension.)

For a two-dimensional representation the row category coordinates may be represented as u_{ik}, $i = 1, 2, \ldots, r$, $k = 1, 2$ and the column category coordinates as $v_{jk}, j = 1, 2, \ldots, c, k = 1, 2$. It can be shown that a large positive residual corresponds to row and column coordinates which are large and of the same sign, a large negative residual to row and column coordinates which are large and of opposite signs, and a small residual to small coordinate values or to coordinates whose signs are not consistent for $k = 1$ and 2. A number of examples will clarify the technique.

The data in Table 3.9 show the hair colour and eye colour of a large number of people. The overall chi-square statistic for

Table 3.9 *Hair colour and eye colour data*

			Hair colour		
Eye colour	Fair	Red	Medium	Dark	Black
Light	688	116	584	188	4
Blue	326	38	241	110	3
Medium	343	84	909	412	26
Dark	98	48	403	681	81

Table 3.10   *Results from applying correspondence analysis to hair colour, eye colour data*

Eye colour	u_1	u_2	Hair colour	v_1	v_2
Light	− 0.535	− 0.276	Fair	− 0.633	− 0.521
Blue	− 0.327	− 0.348	Red	− 0.120	− 0.064
Medium	0.043	0.810	Medium	− 0.059	0.756
Dark	0.778	− 0.381	Dark	0.670	− 0.304
			Black	0.362	− 0.245
	δ_1	δ_2	δ_3	δ_4	
Eigenvalues	1073.3	162.12	4.6	0.0	

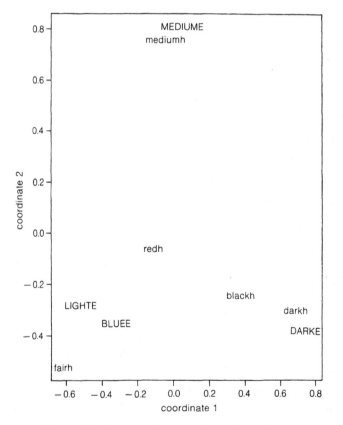

Figure 3.1   *Correspondence analysis of eye data.*

Table 3.11 *Murder data*

Method	1970	1971	1972	1973	1974	1975	1976	1977
Shooting and explosion	15	15	31	17	42	49	38	27
Stabbing	95	113	94	125	124	126	148	127
Blunt instrument	23	16	34	34	35	33	41	41
Poison	9	4	8	3	5	3	1	4
Manual violence	47	60	54	70	69	66	70	60
Strangulation	43	45	43	53	51	63	47	51
Smothering or drowning	26	16	20	24	15	15	15	15

independence is very large since clearly the two variables are not independent. To display graphically the association between hair colour and eye colour a correspondence analysis was performed, leading to the two-dimensional coordinates for row and column categories shown in Table 3.10. These are plotted in Figure 3.1. (The first two coordinates account for 99% of the inertia and therefore give a very accurate representation of the hair colour/eye colour relationship.) The interpretation of this diagram is very clear. For example, the points representing dark eyes and dark hair are close to each other and some way from the origin, indicating a large positive residual for this cell. Dark eyes and fair hair are represented by points on opposite sides of the origin and both are some distance from the origin; consequently the corresponding cell in the table will have a large negative residual.

To illustrate this method of analysis further the data shown in Table 3.11 will be used. These data show the methods by which victims of persons indicted for murder were killed between 1970 and

Table 3.12

	Row coordinates	
Shooting and explosion	− 0.29	− 0.15
Stabbing	0.01	0.06
Blunt instrument	− 0.07	− 0.05
Poison	0.44	− 0.46
Manual violence	0.02	0.04
Strangulation	0.03	− 0.01
Smothering or drowning	0.30	− 0.10
	Column coordinates	
1970	0.25	− 0.10
1971	0.12	0.12
1972	0.2	− 0.16
1973	0.11	0.09
1974	− 0.10	− 0.04
1975	− 0.15	− 0.05
1976	− 0.1	0.08
1977	− 0.03	0.03

The eigenvalues were as follows:

Axis	Eigenvalue	% of inertia
1	0.0166	54.7
2	0.0081	26.6
3	0.0035	11.7
4	0.0010	3.4

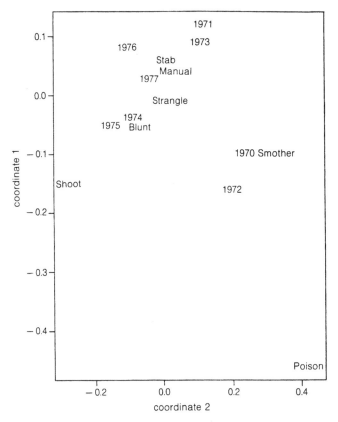

Figure 3.2   *Correspondence analysis of murder data.*

1977. The first two coordinates for each row and column found from a correspondence analysis are given in Table 3.12. (The first two coordinates account for over 80% of the inertia.) A plot of these coordinates is shown in Figure 3.2. The diagram illustrates that the profiles of the methods used for years 1970 and 1972 are similar to each other but rather different from the remaining years. In 1970 smothering was used more than if year and method were independent. Amongst the methods, poison clearly has a very different 1970–77 profile than the others. Shooting also appears to be different, with an obvious 'hump' in the years 1974, 1975 and 1976.

Further interpretation of Figure 3.2 is left to the reader as an exercise.

3.6 Measures of association for contingency tables

In many cases when dealing with contingency tables, a researcher may be interested in **indexing** the strength of the association between the two qualitative variables involved, rather than in simply investigating its significance by means of the usual chi-square test. The purpose may be to compare the degree of association in different tables or to compare the results with others obtained previously.

Many measures of association for contingency tables have been suggested, none of which appears to be completely satisfactory. Several of these measures are based upon the X^2 statistic, which cannot itself be used since its value depends on N, the sample size; consequently it may not be comparable for different tables. A further series of measures, suggested by Goodman and Kruskal (1954), arise from considering the **predictive ability** of one of the variables for the other. Other measures of association have been specifically designed for tables having variables with ordered categories (Chapter 7). In this section a brief description only is given of some of the suggested measure, beginning with those based on the X^2 statistic. A detailed account of measures of association and their properties such as sampling variance, etc., is available in Kendall and Stuart (vol. 2) and in the series of papers by Goodman and Kruskal (1954, 1959, 1963, 1972).

3.6.1 *Measures of association based on the X^2 statistic*

Several traditional measures of association are based upon the standard chi-square statistic, which itself is not a convenient measure since its magnitude depends on N, and increases with increasing N. The simplest way to overcome this is to divide the value of X^2 by N to given what is generally known as the **mean square contingency coefficient** denoted by ϕ^2:

$$\phi^2 = X^2/N. \tag{3.10}$$

However, most researchers are generally happier with measures of association that range between -1 and $+1$ (analogous to the correlation coefficient) or between 0 and 1, with zero indicating independence and unity 'complete association'; consequently ϕ^2 is not very satisfactory since it does not necessarily have an upper limit of 1. A variation of this measure, suggested by Pearson (1904) and called the **coefficient of contingency** is given by:

$$P = \sqrt{\left(\frac{X^2/N}{1 + X^2/N}\right)}. \tag{3.11}$$

This coefficient clearly lies between 0 and 1 as required, and attains its lower limit in the case of complete independence, that is when $X^2 = 0$. In general, however, P cannot attain its upper limit, and Kendall and Stuart (Vol. 2, Ch. 33) show that, even in the case of complete association, the value of P depends on the number of rows and columns in the table. To remedy this the following function of X^2 has been suggested:

$$T = \frac{X^2/N}{\sqrt{[(r-1)(c-1)]}}. \tag{3.12}$$

This again takes the value 0 in the case of complete independence, and, as shown by Kendall and Stuart, may attain a value of $+1$ in the case of complete association when $r = c$ but cannot do so if $r \neq c$. A further modification suggested by Cramer (1946), which may attain the value $+1$ for all values of r and c in the case of complete association, is as follows:

$$C = \frac{X^2/N}{\min(r-1, c-1)}. \tag{3.13}$$

When the table is square, that is when $r = c$, then $C = T$, but otherwise $C > T$, although the difference between them will not be large unless r and c are very different.

The standard errors of all these coefficients can be deduced from the standard error of the X^2 statistic and are given in Kendall and Stuart (Vol. 2).

The major problem with all the above measures of association is that they have no obvious probabilistic interpretation in the same way as has, for example, the correlation coefficient. Interpretation of obtained values is therefore difficult. This has led Goodman and Kruskal (op. cit). to suggest a number of coefficients that are readily interpretable in a **predictive** sense, and these are now briefly described.

3.6.2   Goodman and Kruskal's lambda measures

Goodman and Kruskal in their 1954 paper describe several measures of association that are useful in the situation where the two variables involved cannot be assumed to have any relevant underlying continua and where there is no natural ordering of interest. The rationale behind these measures is the question: 'How much does a knowledge of the classification of one of the variables improve one's ability to predict the classification on the other variable?' Now investigators might claim that their interest lies simply in the 'relationship' between the variables and not in predicting one from another. However, it is difficult to discuss the meaning of an association between two variables without discussing the degree to which one is predictable from the other and the accuracy of the prediction. It seems reasonable therefore to incorporate this notion of predictability into the formal requirements of any index that purports to measure the degree of association between two variables.

To introduce the measures suggested by Goodman and Kruskal, the data in Table 3.13 which shows the frequencies obtained when 284 consecutive admissions to a psychiatric hospital are classified with respect to social class and diagnosis, will be used. Assume that for these data interest centres on the extent to which knowledge of a patient's social class is useful in predicting their diagnostic category.

First suppose that a patient is selected at random and you are asked to guess the diagnosis knowing nothing about the social class. On the basis of the marginal totals for diagnosis in Table 3.13 the best guess would be 'depressed' since this is the diagnosis with the largest marginal total; consequently the probability of the guess being in error would be given by:

$P_1 = P$ (error in guessing diagnosis when social class is unknown)

$\quad = P$ (patient is neurotic) $+ P$ (patient has personality disorder)

$\quad\quad + P$ (patient is schizophrenic)

$\quad = 1 - P$ (patient is depressed),

Table 3.13 *Social class and diagnostic category for a sample of psychiatric patients*

		Neurotic	Depressed	Personality disorder	Schizophrenic	
			Diagnosis (variable B)			
Social class	1	45	25	21	18	109
(variable A)	2	10	45	24	22	101
	3	17	21	18	18	74
		72	91	63	58	284

and therefore

$$P_1 = 1 - 91/284 = 0.68.$$

Now suppose that a patient is again selected at random and again it is required to guess the diagnosis, but in this instance the social class is known. Here the best guess would be the diagnostic category with the largest frequency in the particular social class involved, that is for social class 1, neurotic, and for social classes 2 and 3, depressed. The probability of the guess being in error for each of the three social classes may be obtained as follows:

$p_1 = P$ (error in guessing diagnosis when told patient is in social class 1)

$\quad = 1 - P$ (neurotic is social class 1)

$\quad = 1 - 45/109 = 0.59.$

$p_2 = P$ (error in guessing diagnosis when told patient is in social class 2)

$\quad = 1 - P$ (depressed in social class 2)

$\quad = 1 - 45/101 = 0.55.$

$p_3 = P$ (error in guessing diagnosis when told patient is in social class 3)

$\quad = 1 - P$ (depressed in social class 3)

$\quad = 1 - 21/74 = 0.72.$

The overall probability of an error in guessing the diagnosis of a patient when told their social class may now be obtained as follows:

$P_2 = P$ (error in guessing diagnosis when told social class)

$\quad = p_1 P$ (patient is from social class 1)

$\qquad + p_2 P$ (patient is from social class 2)

$\qquad + p_3 P$ (patient is from social class 3)

$$= 0.59 \times \frac{109}{284} + 0.55 \times \frac{101}{294} + 0.72 \times \frac{74}{284}$$

$$= 0.61.$$

Consequently knowledge of a patient's social class reduces to some degree the probability of an error in predicting their diagnostic category. Goodman and Kruskal's index of predictive ability, λ_B, is computed from these probabilities as follows:

$$\lambda_B = \frac{P_1 - P_2}{P_1} \tag{3.14}$$

$$= \frac{0.68 - 0.61}{0.68} = 0.103.$$

λ_B is the relative decrease in the probability of an error in guessing diagnosis as between social class unknown and known. As such it is readily interpretable. In this example, for instance, it is possible to say that, in prediction of diagnosis from social class, information about social class reduces the probability of error by some 10% on average, an amount that would be unlikely to have any practical significance. In general λ_B may be calculated as follows:

$$\lambda_B = \frac{\sum_{i=1}^{r} \max_j(n_{ij}) - \max_j(n_{.j})}{N - \max_j(n_{.j})}. \tag{3.15}$$

It is, of course, entirely possible to reverse the roles of variables A and B and obtain the index, λ_A, which is suitable for predictions of A from B. λ_A would be calculated as follows:

$$\lambda_A = \frac{\sum_{j=1}^{c} \max_i(n_{ij}) - \max_i(n_{i.})}{N - \max_i(n_{.i})}. \tag{3.16}$$

In general the two indices λ_A and λ_B will be different since situations may arise where B is predictable from A, but not from B, and vice versa.

The value of λ_B (and, of course λ_A) ranges between 0 and 1. If the information about the predictor variable does not reduce the probability of making an error in guessing the category of the other variable, the index is zero, and it may be concluded that there is no predictive association between the two variables. On the other hand, if the index is unity, no error is made, given knowledge of the predictor variable, and consequently there is complete predictive association.

The coefficients λ_B and λ_A are specifically designed for the asymmetric situation in which explanatory and dependent variables are clearly defined. The same 'reduction in error' approach can be used to produce a coefficient for the symmetric situation where neither variable is specially designated as that to be predicted. Instead it is supposed that sometimes one and sometimes the other variable is given beforehand and that the one not given has to be predicted.

This coefficient λ is given by:

$$\lambda = \frac{\sum_{i=1}^{r} \max_j(n_{ij}) + \sum_{j=1}^{c} \max_i(n_{ij}) - \max_j(n_{.j}) - \max_i(n_{i.})}{2N - \max_j(n_{.j}) - \max_i(n_{i.})}.$$

$$(3.17)$$

For the data in Table 3.13 this leads to

$$\lambda = \frac{(45 + 45 + 21) + (45 + 45 + 24 + 22) - 91 - 109}{2 \times 284 - 91 - 109}$$

$$= 0.128.$$

The coefficient shows the relative reduction in the probability of an error in guessing the category of either variable as between knowing and not knowing the category of the other. λ will always take a value between that of λ_A and λ_B.

A problem arises in the use of lambda measures of association when the marginal distributions are far from being uniform. In such cases the values of the indices may be misleadingly low and with extremely skewed marginal distributions it appears that any measure applied to the raw data may be inappropriate. Other problems associated with these measures are discussed in the papers of Goodman and Kruskal previously referenced.

Potential users of these measures should remember that most are specifically designed for particular types of situation and that the choice of the most appropriate measure depends critically on careful consideration of the type of data involved. Calculating any measure without regard to its suitability for the particular data set under investigation would clearly not be sensible.

3.7 Summary

In this chapter the analysis of the general $r \times c$ contingency table has been discussed. Many investigators, having arrived at a significant chi-square value for such a table, would often proceed no further. In general, however, more detailed investigation of the reasons for the significant association is needed using the methods described in the previous sections such as the calculation of residuals or the application of correspondence analysis.

CHAPTER 4

Multidimensional tables

4.1 Introduction

The methods of analysis described in the previous two chapters involved *two-dimensional* contingency tables arising from observations made on two categorical variables. In many situations, however, tables of counts resulting from the cross-classification of more than two categorical variables are of interest. For example, Table 4.1 shows a three-dimensional contingency table concerned with suicide behaviour and Table 4.2 a four-dimensional table describing the voting intention of a sample of individuals.

The analysis of three-dimensional tables poses entirely new conceptual problems as compared with the analysis of those of two dimensions. However, the extension from tables of three dimensions to those of four or more, whilst often increasing the complexity of both analysis and interpretation, presents no further new problems; consequently the introduction given to these higher dimensional tables in this chapter will be in terms of those arising from three categorical variables. This will form the basis of the more detailed coverage of the topic to be given in Chapter 5, where the fitting of models to contingency tables is described.

Much work has been done on the analysis of multidimensional contingency tables, particularly during the last two decades. Lewis (1962) gives an excellent review, and a selection of the large number of other relevant references are those of Darroch (1962), Birch (1963), Bishop (1969), Fienberg (1970), Goodman (1968, 1970, 1971), Bishop *et al.* (1975), Fienberg (1987), Upton (1986), Freeman (1987) and Agresti (1990).

4.2 Nomenclature for three-dimensional tables

The nomenclature used previously for dealing with an $r \times c$ table is easily extended to deal with a three-dimensional $r \times c \times l$

Table 4.1 Suicide behaviour. Age by sex by cause of death

				Causes of Death		
Age group	(1) Solid or liquid matter	(2) Gas	(3) Hanging, strangling, suffocating, drowning	(4) Gun, knives, explosives	(5) Jumping	(6) Other
Male						
10–40 (A1)	398	121	455	155	55	124
40–70 (A2)	399	82	797	168	51	82
> 70 (A3)	93	6	316	33	26	14
Female						
10–40 (A1)	259	15	95	14	40	38
40–70 (A2)	450	13	450	26	71	60
> 70 (A3)	154	5	185	7	38	10

Table 4.2   *Voting behaviour: vote by sex by class by age*

Age group	Men		Women	
	Conservative	Labour	Conservative	Labour
		Upper middle class		
> 73	4	0	10	0
51–73	27	8	26	9
41–50	27	4	25	9
26–40	17	12	28	9
< 26	7	6	7	3
		Lower middle class		
> 73	8	4	9	2
51–73	21	13	33	8
41–50	27	12	29	4
26–40	14	15	17	13
< 26	9	9	13	7
		Working class		
> 73	8	15	17	4
51–73	35	62	52	53
41–50	29	75	32	70
26–40	32	66	36	67
< 26	14	34	18	33

Reproduced from Payne (1977) by permission of John Wiley and Sons.

contingency table having r rows, c columns and l 'layer' categories. The observed frequency in the ijkth cell is now represented by n_{ijk} for $i = 1, 2, \ldots, r, j = 1, 2, \ldots, c$ and $k = 1, 2, \ldots, l$. By summing the n_{ijk} over different subscripts various marginal totals may be obtained. Summing over all values of both i and j, for example, will yield the total for the kth layer category. Similarly the totals for the ith row categories and the jth column category are obtained by summing the n_{ijk} over j and k, and over i and k respectively. These totals are known as **single variable marginals**:

$$\left.\begin{aligned}
n_{i..} &= \sum_{j=1}^{c} \sum_{k=1}^{l} n_{ijk}, \\
n_{.j.} &= \sum_{i=1}^{r} \sum_{k=1}^{l} n_{ijk}, \\
n_{..k} &= \sum_{i=1}^{r} \sum_{j=1}^{c} n_{ijk}.
\end{aligned}\right\} \qquad (4.1)$$

For example, for the data shown in Table 4.1, these single variable marginal totals are as follows:

Number of men $= n_{1..} = 398 + 121 + \cdots + 14 = 3375$.
Number of women $= n_{2..} = 259 + 15 + \cdots + 10 = 1930$.
Number of deaths by method $1 = n_{.1.} = 398 + 399 + \cdots + 154$
$$= 1753.$$
Number of deaths by method $2 = n_{.2.} = 121 + 82 + \cdots + 5 = 242$.
Similarly $n_{.3.} = 2298$, $n_{.4.} = 403$, $n_{.5.} = 281$, $n_{.6.} = 328$.
Number of deaths in age group 10–40 $= n_{..1} = 398 + 121 + \cdots + 124$
$$+ 259 + 15$$
$$+ \cdots + 38 = 1769.$$

Similarly $n_{..2} = 2649$, $n_{..3} = 887$

Summing the n_{ijk} over any single subscript gives the two-variable marginal totals:

$$\left. \begin{aligned} n_{ij.} &= \sum_{k=1}^{l} n_{ijk}, \\ n_{i.k} &= \sum_{j=1}^{c} n_{ijk}, \\ n_{.jk} &= \sum_{i=1}^{r} n_{ijk}. \end{aligned} \right\} \qquad (4.2)$$

For example, Table 4.3 shows the two-variable marginals for the data of Table 4.1, obtained by summing over the third variable, age. Similar tables could be obtained by summing over either of the other two variables. The grand total of the observed frequencies, $n_{...}$ is given by:

$$n_{...} = \sum_{i=1}^{r} \sum_{j=1}^{c} \sum_{k=1}^{l} n_{ijk}, \qquad (4.3)$$

and as usual denoted by N.

Table 4.3 *Two variable marginal totals for suicide data obtained by summing over age*

| | Method | | | | | |
	1	2	3	4	5	6
Male	890	209	1568	356	132	220
Female	863	33	730	47	149	108

This nomenclature is easily generalized to contingency tables involving more than three variables. (A similar nomenclature is used for the population probabilities, p_{ijk}, and the estimated expected values, E_{ijk}.)

4.3   Why analyse multidimensional tables?

Researchers with data in the form of a multidimensional contingency table may ask why they should not simply attempt its analysis by examining all the two-dimensional tables arrived at by summing over the other variables. Reasons why this would not, in most cases, be an appropriate procedure are not difficult to find. The most compelling is that it can lead to very misleading conclusions being drawn about the data. Why this should be so will become clear after the discussion later of such concepts as partial and conditional independence. Here it will suffice to illustrate the problem by means of an example using the data shown in Table 4.4 taken from Bishop (1969).

Analysing first only the data for clinic A, the chi-squared statistic is found to be almost zero. Similarly for the data from clinic B, chi-squared is approximately zero. If, however, the data are collapsed over clinics, the chi-square becomes 5.26 which with one d.f. is significant beyond the 5% level; consideration only of this collapsed table would therefore lead to the erroneous conclusion that survival and amount of pre-natal care are related. The reason for such spurious results will be made clear later. This example should, however, make it clear why consideration of all two-dimensional tables is not a sufficient procedure for the analysis of multi-dimensional tables.

The analysis of multidimensional tables presents problems not

Table 4.4   *Three-dimensional contingency table relating survival of infants to amount of pre-natal care received in two clinics*

		Died		Survived	
		Less	*More*	*Less*	*More*
Amount of pre-natal care					
Place where	Clinic A	3	4	176	293
care received	Clinic B	17	2	197	23

met for two-dimensional tables, where a single hypothesis, namely that of the independence of the two variables involved, is of interest. In the case of multidimensional tables, more than one hypothesis may be of concern. For example, an investigator may wish to test that some variables are independent of some others, or that a particular variable is independent of the remainder. The simplest hypothesis of interest for a multidimensional table is, however, that of the mutual independence of the variables; in the following section testing such a hypothesis for a three-dimensional table is considered.

4.4 Testing the mutual independence of the variables in a three-way table

The hypothesis of the mutual independence of the variables in a three-dimensional contingency table may be formulated as follows:

$$H_0 : p_{ijk} = p_{i..}p_{.j.}p_{..k}, \tag{4.4}$$

where p_{ijk} represents the probability of an observation being in the ijkth cell of the table, and $p_{i..}, p_{.j.}$ and $p_{..k}$ are the marginal probabilities of the row column and layer variables respectively. This is the three-dimensional equivalent of the hypothesis of independence in a two-way table (equation (1.6)). To test this hypothesis an exactly analogous procedure to that for the two-variable case is used (section 1.5). Firstly, estimates of the frequencies to be expected when H_0 is true are calculated. Next, these estimated expected values are compared with the observed frequencies by means of the usual chi-square statistic. Lastly the number of degrees of freedom of the test statistic is found and the significance level of the statistic assessed. In the case of the hypothesis of the mutual independence of the three variables the expected values may be obtained in a similar way to that used for two-way tables, i.e.:

$$E_{ijk} = N \hat{p}_{i..}\hat{p}_{.j.}\hat{p}_{..k}, \tag{4.5}$$

(cf. equation (1.9)) where $\hat{p}_{i..}, \hat{p}_{.j.}$ and $\hat{p}_{..k}$ are estimates of the corresponding probabilities. It is easy to show that the best estimates are derived from the relevant single variable marginal totals, namely:

$$\hat{p}_{i..} = \frac{n_{i..}}{N}, \hat{p}_{.j.} = \frac{n_{.j.}}{N}, \hat{p}_{..k} = \frac{n_{..k}}{N}. \tag{4.6}$$

(Again as in Chapter 1, these are maximum likelihood estimates.)

Substituting these values in (4.5) gives:

$$E_{ijk} = N \frac{n_{i..}}{N} \frac{n_{.j.}}{N} \frac{n_{..k}}{N}$$

$$= \frac{n_{i..} n_{.j.} n_{..k}}{N^2}. \tag{4.7}$$

Having obtained the expected values using (4.7) the test statistic is calculated in the usual way as:

$$X^2 = \sum_{i=1}^{r} \sum_{j=1}^{c} \sum_{k=1}^{l} \frac{(n_{ijk} - E_{ijk})^2}{E_{ijk}}. \tag{4.8}$$

To complete the procedure the degrees of freedom are needed. In the case where the hypothesis is that of the mutual independence of the three variables, the appropriate value is given by:

$$\text{d.f.} = rcl - r - c - l + 2. \tag{4.9}$$

In the case of other hypotheses which might be of interest (see following section) the value for the degrees of freedom will depend upon the particular hypothesis being tested. A general procedure for determining degrees of freedom for chi-square tests on multi-dimensional tables is discussed in Section 4.7.

4.4.1 Numerical example

To illustrate the test of mutual independence it will be applied to the data in Table 4.1. The first stage in testing the hypothesis involves calculating expected values using (4.7). For example, the expected value for the male, method 1, age group 10–40, cell E_{111}, is given by:

$$E_{111} = \frac{3375 \times 1769 \times 1753}{5305 \times 5305} = 371.89,$$

and the full set of expected values under the hypothesis of mutual independence is given in Table 4.5. The test statistic calculated from (4.8) is $X^2 = 790.30$ with degrees of freedom from (4.9) equal to 27. Clearly the three variables forming Table 4.1 are not mutually independent. More detailed examination of the data is now needed to assess why the hypothesis is rejected; this is discussed in the next section.

Examining Table 4.5 it is seen that summing the expected values

Table 4.5   *Expected values under mutual independence for suicide data*

| | Method | | | | | |
	1	2	3	4	5	6
Male						
A1	371.89	51.34	487.5	85.49	59.61	69.58
A2	556.89	76.88	730.02	128.02	89.27	104.20
A3	186.47	25.74	244.44	42.87	29.89	34.89
Female						
A1	212.66	29.36	278.78	48.89	34.09	39.79
A2	318.46	43.96	417.46	73.21	51.05	59.59
A3	106.63	14.72	139.78	24.51	17.09	19.95

over any two of the variables gives a total equal to the corresponding single variable marginal total of observed values. For example:

$$E_{1..} = 371.89 + 556.89 + \cdots + 34.89$$
$$= 3375 = n_{1..}.$$

Summing these values over any **single** variable, however, does not give totals equal to the two-variable marginals of the observed values. For example:

$$E_{11.} = 371.89 + 556.89 + 186.47$$
$$= 1115.25 \neq n_{11.} (n_{11.} = 890).$$

The constraints on the marginal totals of expected values in the case of the hypothesis of mutual independence are such that only their single variable marginals, namely $E_{i..}$, $E_{.j.}$ and $E_{..k}$ are required to equal the corresponding marginals of observed values, $n_{i..}, n_{.j.}$ and $n_{..k}$. In the case of other hypotheses various other marginal totals of expected values may be likewise constrained as will be seen in the next section. Such constraints arise from the form of the maximum likelihood equations from which the estimates of expected values are derived; for details see Birch (1963).

4.5   Further hypotheses of interest in three-way tables

If the test of mutual independence described in the previous section gives a non-significant result then further analysis of the table is unlikely to be worth while. When the test gives a significant result, however, it should not be assumed that there are significant associations between all variables. It might be the case, for example, that an association exists between two of the variables whilst the third is completely independent. In this case hypotheses of **partial independence** would be of interest. Again, situations arise where two of the variables are independent in each level of the third, but each may be associated with this third variable. In other words, the first two variables are **conditionally independent** given the level of the third. Such hypotheses may again be formulated in terms of probabilities; for example, the three possible hypotheses of partial independence in a three-dimensional table are:

(1) $H_0^{(1)}: p_{ijk} = p_{i..} p_{.jk}$ (row classification independent of column and layer classification);
(2) $H_0^{(2)}: p_{ijk} = p_{.j.} p_{i.k}$ (column classification independent of row and layer classification);
(3) $H_0^{(3)}: p_{ijk} = p_{..k} p_{ij.}$ (layer classification independent on row and column classification).

Taking the first of these in more detail, the hypothesis states that the probability of an observation occurring in the ijkth cell, that is p_{ijk}, is given by the product of the probability of it falling in the ith category of the row variable, $p_{i..}$, and the probability of its being in the jkth cell of the column × layer classification, $p_{.jk}$. If the hypothesis is true it implies that the row classification is independent of both the column and the layer classification; that is it implies the truth of the following composite hypothesis:

$$p_{ij.} = p_{i..} p_{.j.} \text{ and } p_{i.k} = p_{i..} p_{..k}.$$

To test the hypothesis the procedure is exactly as previously, beginning with the calculation of expected values which in this case are given by:

$$E_{ijk} = N \hat{p}_{i..} \hat{p}_{.jk}, \tag{4.10}$$

with the estimators $\hat{p}_{i..}$ and $\hat{p}_{.jk}$ of the probabilities $p_{i..}$ and $p_{.jk}$ being obtained from the relevant marginal totals as follows:

$$\hat{p}_{i..} = \frac{n_{1..}}{N} \text{(as before)}, \hat{p}_{.jk} = \frac{n_{.jk}}{N}. \tag{4.11}$$

In this case the two-variable marginal totals, $n_{.jk}$, found by summing the observed frequencies over the first variable, are needed. (Again the estimates given by (4.11) are maximum likelihood estimates.)

Using these probability estimates in (4.10) gives:

$$E_{ijk} = N \frac{n_{i..}}{N} \frac{n_{.jk}}{N}$$

$$= \frac{n_{i..} \, n_{.jk}}{N}. \tag{4.12}$$

The test statistic is now calculated again using (4.8) and for this hypothesis has degrees of freedom given by

$$\text{d.f.} = clr - cl - r + 1. \tag{4.13}$$

4.5.1 Numerical example

For the data in Table 4.1 the hypothesis $H_0^{(1)}$ states that method of suicide is independent of sex, and that age and sex are likewise not associated. To calculate the expected values for this hypothesis (4.12) is used. For example:

$$E_{111} = \frac{3375 \times 657}{5305} = 417.98,$$

and the full set of such values is given in Table 4.6. These expected values lead to a value of 520.41 for the chi-square test statistic with degrees of freedom from (4.13) of 17. Clearly the hypothesis is not acceptable for these data.

Returning to Table 4.6 it is seen in this case, in addition to the single variable marginal totals for expected and observed values being equal, the two-variable marginal totals obtained by summing

Table 4.6 *Expected values under partial independence for the suicide data*

| | Method | | | | | |
	1	2	3	4	5	6
Male						
A1	417.98	86.52	349.91	107.52	60.44	103.06
A2	540.13	60.44	793.33	123.42	77.61	90.34
A3	157.14	6.99	318.73	25.45	40.72	15.27
Female						
A1	239.02	49.48	200.09	61.48	34.56	58.94
A2	308.87	34.56	453.67	70.58	44.38	51.66
A3	89.86	4.00	182.25	14.55	23.28	8.73

over sex are also equal, that is $E_{.jk} = n_{.jk}$. For example:

$$E_{.11} = 417.98 + 239.02 = 657 = n_{.11}.$$

Hypotheses of partial independence fix the single variable and one set of two-variable marginal totals of the expected values to be equal to the corresponding totals of the observed values.

The concept of conditional independence mentioned briefly earlier in this chapter will be considered in more detail in Chapter 5, but it is not difficult to see that the data in Table 4.4 illustrate this type of independence; amount of care and survival are clearly independent in each level of the third variable (place where care received), that is they are independent in clinic A **and** in clinic B.

4.6   Second-order relationship in three-way tables

For multidimensional tables the possibility exists of the presence of a more complex relationship between variables than those considered up to this point. For example, in a three-way table an association between two of the variables may differ in degree or in direction in different categories of the third; consequently a conjoint three-variable relationship would need to be allowed for. Such a relationship would be termed **second order** to differentiate it from the **first order** associations between pairs of variables considered up to this point. Second and higher order associations are best understood in terms of the models to be considered in the following chapter. Roy and Kastenbaum (1956), however, have specifically considered the formulation of the hypothesis of no second-order association between the variables in a three-way table in terms of the probabilities, p_{ijk}. The formulation these workers give is as follows:

$$H_0 : \frac{p_{rcl}p_{ijl}}{p_{icl}p_{rjl}} = \frac{p_{rck}p_{ijk}}{p_{ick}p_{rjk}}, \tag{4.14}$$

$i = 1, \ldots, r-1; \quad j = 1, \ldots, c-1; k = 1, \ldots l-1$. Although it is not difficult to show how (4.14) relates to the hypotheses of no first-order association in a two-dimensional table, details will not be given here. Interested readers should consult Bhapkar and Koch (1968) who show that (4.14) arises naturally when extending the hypothesis of independence between pairs of variables to that of no three-variable association for higher-order tables. Essentially the quantity on the left hand of (4.14) represents a measure of association of the first two variables within the *l*th category of the third, and that on the right-hand side the same measure of association between the first two variables within the *k*th category of the third. The hypothesis states then that this measure of association is the same for all categories of the third variable, or in other words, that the association between variables 1 and 2 does not differ with the level of variable 3. (Of course, since the order of variables is arbitrary, the hypothesis of no second-order association between the three variables implies that the association between any pair of variables is the same at all levels of the remaining variable.)

A complication arises when considering how to test the hypothesis given in (4.14), since under this hypothesis, estimates of expected values cannot be found directly from various marginal totals as in the cases previously considered. Instead these estimates must be obtained **iteratively** by a procedure to be described in the next chapter. Having obtained the estimates, however, the test proceeds in the usual fashion with the calculation of the X^2 statistic. Detailed discussion of testing for second- and higher-order relationships in multidimensional tables will be left until Chapter 5.

4.7 Degrees of freedom

A convenient method of determining the degrees of freedom of the X^2 statistic for multidimensional tables is by use of the following general formula:

d.f. = (number of cells in table − 1) − (number of probabilities estimated from the data for the particular hypothesis being tested).

$$(4.15)$$

For example, consider the hypothesis of mutual independence in a three-way table. First, the number of cells is clearly the product rcl. In testing this hypothesis the probabilities $p_{i..}$ $p_{.j.}$ and $p_{..k}$ need to be estimated for all values of i, j and k using equation (4.6). Since probabilities must sum to unity there are $(r - 1)$ row, $(c - 1)$ column and $(l - 1)$ layer probabilities to estimate; hence the degrees of freedom in this case will be:

$$\text{d.f.} = rcl - (r - 1) - (c - 1) - (l - 1) - 1$$
$$= rcl - r - c - l + 2.$$

Now consider the hypothesis of partial independence discussed in section 4.5. In this case the probabilities $p_{i..}$ and $p_{.jk}$ have to be estimated. As before there are $(r-1)$ row probabilities and the number of column × layer probabilities is simply $(cl-1)$, leading to the following value for degrees of freedom:

$$\text{d.f.} = rcl - (r-1) - (cl-1) - 1$$
$$= rc - r - cl + 1.$$

4.8 Likelihood ratio criterion

An alternative criterion to the usual X^2 statistic for comparing observed frequencies with estimated values of those expected under a particular hypothesis is the **likelihood ratio criterion**, X_L^2, given by:

$$X_L^2 = 2 \sum \text{observed} \times \ln(\text{observed/expected}), \qquad (4.16)$$

which like X^2 has a chi-square distribution when the hypothesis under test is true. (The degrees of freedom of X_L^2 are of course the same as for X^2.) Since X^2 is easily shown to be an approximation to X_L^2 for large samples, the two statistics will take similar values for many tables. However, Ku and Kullback (1974), Williams (1976) and others have shown that for a variety of reasons, X_L^2 is generally preferable to X^2; consequently it will be the goodness-of-fit criterion primarily used in the remainder of the text.

4.9 Summary

In this chapter contingency tables arising from more than two categorical variables have been introduced. Some of the new problems arising from this extension have been considered, in particular the increased number of hypotheses that may be of interest, and the possibility of second and higher order relationships.

CHAPTER 5

Log-linear models for contingency tables

5.1 Introduction

The previous chapters have dealt almost exclusively with **hypothesis testing** techniques for the analysis of contingency tables. In this chapter an alternative approach will be considered, namely that of **fitting models** and **estimating** the parameters in the models. The term model refers to some 'theory' or conceptual framework about the observations, and the parameters in the model represent the 'effects' that particular variables or combinations of variables have in determining the values taken by the observations. Such an approach is common in many branches of statistics such as regression analysis and the analysis of variance. Most common are **linear models** which postulate that the expected values of the observations are given by a linear combination of a number of parameters. Techniques such as maximum likelihood and least squares may be used to **estimate** the parameters, and estimated parameter values may then be used in identifying which variables are of greatest importance in 'predicting' the observed values.

The models for contingency table data to be discussed in the following section are very similar in many respects to those used for quantitative data in, for example, the analysis of variance, and readers not familiar with such models are referred to Hays (1973, Ch. 12). A consequence of this similarity is that an analysis-of-variance term, **interaction**, is now often used as an alternative to the term association when describing a relationship between the qualitative variables forming a contingency table. Such a terminology will be used throughout the remainder of this text, interactions between **pairs** of variables being referred to as **first-order** and those between **triplets** of variables as **second-order** and so on.

The major advantages to be gained from the model-fitting techniques to be described are firstly that they provide a systematic approach to the analysis of complex multidimensional tables and secondly that they provide estimates of the magnitude of effects of interest; consequently they allow the relative importance of different effects to be judged.

5.2   Log-linear models

Returning for the moment to two-dimensional tables, the hypothesis of independence, that is of no first-order interaction between the two variables, specifies, as was seen in Chapter 2, that:

$$p_{ij} = p_{i.} p_{.j}. \qquad (5.1)$$

This equation specifies a particular structure or model for the data, namely that in the population the probability of an observation falling in the ijth cell of the table is simply the **product** of the marginal probabilities. The first question to ask is how this model could be rearranged so that p_{ij} or some function of it can be expressed as the **sum** of the marginal probabilities or some function of them. The model would then begin to correspond to those found in the analysis of variance of quantitative data. By taking the natural logarithms of (5.1) such a rearrangement is easily found, namely:

$$\ln p_{ij} = \ln p_{i.} + \ln p_{.j}. \qquad (5.2)$$

This may be rewritten in terms of theoretical frequencies, $F_{ij}(F_{ij} = N p_{ij}$, etc.; see Chapter 1) as:

$$\ln F_{ij} = \ln F_{i.} + \ln F_{.j} - \ln N. \qquad (5.3)$$

Summing (5.3) over i leads to:

$$\sum_{i=1}^{r} \ln F_{ij} = \sum_{i=1}^{r} \ln F_i + r \ln F_{.j} - r \ln N, \qquad (5.4)$$

and over j:

$$\sum_{j=1}^{c} \ln F_{ij} = c \ln F_{i.} + \sum_{j=1}^{c} \ln F_{.j} - c \ln N, \qquad (5.5)$$

and finally over i and j gives:

$$\sum_{i=1}^{r} \sum_{j=1}^{c} \ln F_{ij} = c \sum_{i=1}^{r} \ln F_{i.} + r \sum_{j=1}^{c} \ln F_{.j} - rc \ln N. \qquad (5.6)$$

It is now a matter of simple algebra to show that equation (5.3) may be rewritten in a form reminiscent of the models used in the analysis of variance, namely:

$$\ln F_{ij} = u + u_{1(i)} + u_{2(j)}, \tag{5.7}$$

where

$$u = \frac{\sum_{i=1}^{r} \sum_{j=1}^{c} \ln F_{ij}}{rc}, \tag{5.8}$$

$$u_{1(i)} = \frac{\sum_{j=1}^{c} \ln F_{ij}}{c} - u, \tag{5.9}$$

$$u_{2(j)} = \frac{\sum_{i=1}^{r} \ln F_{ij}}{r} - u. \tag{5.10}$$

Equation (5.7) specifies a linear model for the logarithms of the frequencies, or, in other words, what is generally known as a **log-linear** model. Its similarity to the models used in the analysis of variance is clear; consequently analysis of variance terms are used for the parameters, and u is said to represent an 'overall mean effect', $u_{1(i)}$ represents the 'main effect' of the ith category of variable 1 and $u_{2(j)}$ the 'main effect' of the jth category of variable 2. Examining equations (5.9) and (5.10) it is seen that the main effect parameters are measured as deviations of row or column means of log-frequencies from the overall mean so that

$$\sum_{i=1}^{r} u_{1(i)} = 0, \qquad \sum_{j=1}^{c} u_{2(j)} = 0,$$

or using an obvious dot notation:

$$u_{1(.)} = 0, \qquad u_{2(.)} = 0.$$

The above derivation of the model specified in (5.7) has been in terms of the theoretical frequencies, F_{ij}. In practice of course these and the parameters in the model will need to be estimated, and the adequacy of the proposed model for the observed data tested. Details of fitting log-linear models to contingency table data are given in section 5.3.

Also in section 5.3 it will be shown that the values taken by the 'main effect' parameters simply reflect differences between the row or the column marginal total and so in the context of contingency table analysis are of little concern (in contrast to the analysis of variance situation where main effects are often of major importance). What is of interest, however, is to extend the model specified in (5.7) to the situation in which the variables cannot be assumed to be independent. To do this, an extra term representing the interaction between the variables is introduced into the model giving:

$$\ln F_{ij} = u + u_{1(i)} + u_{2(j)} + u_{12(ij)}. \tag{5.11}$$

The reason for the nomenclature used for the parameters now becomes clear. The numerical subscripts of the parameters denote the particular variables involved, and the alphabetical subscripts the categories of those variables in the same order. Thus $u_{12(ij)}$ represents the interaction effect between levels i and j of variables 1 and 2 respectively. As will be shown in the next section interaction effects are again measured as deviations so that:

$$\sum_{j=1}^{c} u_{12(ij)} = 0, \qquad \sum_{i=1}^{r} u_{12(ij)} = 0,$$

that is

$$u_{12(i.)} = 0, \qquad u_{12(.j)} = 0.$$

Estimation of the interaction effects would now be useful in identifying those categories responsible for any departure from independence. In terms of the interaction parameters the hypothesis of independence is equivalent to requiring that $u_{12(ij)} = 0$ for all values of i and j. Testing for independence is seen therefore to be equivalent to testing whether all the interaction terms in (5.11) are zero or, equivalently, that the model specified in (5.7) provides an adequate fit to the data. It should be noted that the model in (5.11) will fit the data perfectly since the expected values under this model are simply the observed frequencies. This is so because the number of parameters in the model is equal to the number of cell frequencies. For this reason (5.11) is known as the **saturated model** for a two-dimensional table. The use of saturated models when analysing complex multidimensional contingency tables is considered in section 5.7.

For a three-dimensional table the saturated log-linear model can be written as:

$$\ln F_{ijk} = u + u_{1(i)} + u_{2(j)} + u_{3(k)} + u_{12(ij)} + u_{13(ik)} + u_{123(ijk)}. \tag{5.12}$$

This model includes main effect parameters for each variable ($u_{1(i)}$, $u_{2(j)}$ and $u_{3(k)}$), first-order interaction effect parameters for each pair of variables ($u_{12(ij)}$, $u_{13(ik)}$ and $u_{23(jk)}$), and, in addition, parameters representing possible second-order effects between the three variables

($u_{123(ijk)}$). The models corresponding to the various hypotheses connected with such a table discussed previously in Chapter 4 are obtained by equating certain of the terms in (5.12) to zero. Before considering these models in detail, however, mention must be made that in this text attention is restricted to what Bishop terms **hierarchical models**. These are such that whenever a higher-order effect is included in the model, the lower-order effects composed from variables in the higher effect are also included. So, for example, if a term u_{123} is included in a model, terms u_{12}, u_{13}, u_{23}, u_1, u_2 and u_3 **must** also be included. Consequently models such as

$$\ln F_{ijk} = u + u_{2(j)} + u_{3(k)} + u_{123(ijk)} \tag{5.13}$$

are not permissible.

The restriction to hierarchical models arises from the constraints imposed by the maximum likelihood estimation procedures, details of which are outside the scope of this text. In practice the restriction to hierarchical models is of little consequence since most tables can be described by a series of such models, although this may, in some cases, require the table to be partitioned in some way (see section 5.7).

Returning now to the model specified by (5.12), it is of interest to consider how the hypothesis of no second-order interaction, previously expressed in terms of probabilities, p_{ijk} (see equation (4.14)), may be expressed in terms of the u-parameters. It is relatively easy to show that the model equivalent of (4.14) is achieved by setting the parameters $u_{123(ijk)}$ equal to zero for all values of i, j and k, so that the hypothesis becomes (taking the alphabetical subscripts as understood):

$$H_0 : u_{123} = 0. \tag{5.14}$$

Similarly other hypotheses of interest may also be expressed in terms of the parameters in (5.12). The hypothesis of mutual independence, for example (Equation (4.4)), which specifies that there are no associations of any kind between the three variables or, in other words, that there are no first-order interactions between any pair of variables, and no conjoint three-variable interaction, may now be written as follows:

$$H_0 : u_{12} = 0, u_{13} = 0, u_{23} = 0, u_{123} = 0. \tag{5.15}$$

In this case the model for log-frequencies is simply:

$$\ln F_{ijk} = u + u_{1(i)} + u_{2(j)} + u_{3(k)}, \tag{5.16}$$

involving only an overall mean effect and main effect parameters for each of the three variables. If such a model provides an adequate fit to the data it implies that differences between cell frequencies simply reflect differences between single variable marginal totals.

Now consider a model which specifies that $u_{12} = 0$ and therefore necessarily for hierarchical models has $u_{123} = 0$. The log-linear model would now be of the form:

$$\ln F_{ijk} = u + u_{1(i)} + u_{2(j)} + u_{3(k)} + u_{13(ik)} + u_{23(jk)}. \qquad (5.17)$$

Setting $u_{123} = 0$ is equivalent to postulating that the interaction between variables 1 and 2 is the same at all levels of variable 3; setting $u_{12} = 0$ is equivalent to postulating that this interaction is zero. Consequently the model in (5.17) is seen to be specifying that there is no interaction between variables 1 and 2 at each level of variable 3, or in other words, that variables 1 and 2 are conditionally independent given variable 3. In such a model, variables 1 and 2 are each assumed to be associated with variable 3 since neither u_{13} nor u_{23} has been set to zero.

Similarly the hypothesis of partial independence discussed in the preceding chapter can be considered in terms of (5.12) with some parameters set equal to zero. In this case u_{123} and one pair of u_{12}, u_{13} and u_{23} would be specified to be zero. Hypothesis $H_0^{(1)}$ (section 4.5), for example, is equivalent to the following log-linear model:

$$\ln F_{ijk} = u + u_{1(i)} + u_{2(j)} + u_{3(k)} + u_{23(jk)}. \qquad (5.18)$$

If u terms continue to be deleted from (5.12) so that there are fewer terms than in the complete independence model, (5.16), models are obtained which do not include all variables. These are termed **non-comprehensive** models by Bishop et al.; if such a model is tenable for a set of data it simply implies that one (or more) of the variables is redundant and that the dimensionality of the table could be reduced accordingly. In practice, of course, only **comprehensive** models, that is those containing at least a main effect parameter for each variable, would be of concern.

Bishop et al. prove that a three-dimensional contingency table may be collapsed over any variable that is independent of at least one of the remaining pair, and the reduced table examined without the danger of misleading conclusions previously alluded to (section 4.3). Such a result shows that acceptance of one of the hypotheses of partial independence, that is showing that a model such as (5.18) provides an adequate fit to the data, would allow the table to be collapsed over **any** of the three variables with a consequent simplification in the analysis. The result also shows that where only conditional independence holds (that is the model specified in (5.17)), care must be taken in deciding which variables may be collapsed over. For example, the spurious result found when collapsing Table 4.4 over the clinic variable is now explained, since **each** of the remaining two variables, amount of pre-natal care and survival, is associated with the 'clinic where care received' variable.

5.3 Fitting log-linear models and estimating parameters

In the preceding section the equivalence of particular log-linear models to particular hypotheses about multidimensional contingency tables was illustrated. Consequently assessing the adequacy of a suggested model for the data follows exactly the same lines as used in the hypothesis-testing approach; estimates of the frequencies to be expected if the model is correct are calculated and these are compared in the usual way to the observed values using either the X^2 or X_L^2 statistic. The estimated expected values are obtained as for the corresponding hypothesis and, as mentioned in Chapter 4, may in some cases be calculated explicitly from the relevant marginals of the observed table, but in other cases must be obtained using the type of iterative procedure to be discussed in section 5.5.

An advantage of the fitting of log-linear models is that estimates of parameters may be obtained which are often useful in quantifying the effects of various variables and of interactions between variables. Estimates of the parameters in the fitted model are obtained as functions of the logarithms of the E_{ijk} and the form of such estimates is very similar to those used for the parameters in analysis of variance models as the following examples will illustrate. Examining first the two-dimensional table shown in Table 5.1, estimates of main effect parameters in the model

$$\ln F_{ij} = u + u_{1(i)} + u_{2(j)} \qquad (5.19)$$

will be found.

To begin, the expected values under (5.19) are needed. Here the model is equivalent to the hypothesis of the independence of the two variables so that required values are found from (1.9) and are given in Table 5.2. Estimates of the main effect parameters are now found by simply substituting the values from Table 5.2 for the F_{ij}

Table 5.1   *Two-dimensional data*

		Variable 2			
		1	2	3	
	1	20	56	24	100
Variable 1	2	8	28	14	50
	3	2	16	2	20
		30	100	40	170

Table 5.2   *Expected values for data of Table 5.1*

		Variable 2			
		1	2	3	
	1	17.65	58.82	23.53	100
Variable 1	2	8.82	29.41	11.76	50
	3	3.53	11.76	4.71	20
		30	100	40	170

in (5.9) and (5.10). For example:

$$\hat{u}_{11} = \frac{1}{3}(\ln 17.65 + \ln 58.82 + \ln 23.53) - \frac{1}{9}(\ln 17.65 + \cdots + \ln 4.71)$$

$$= 0.77.$$

The estimated main effects are shown in Table 5.3. Note first that the estimates for each variable sum to zero so that the last effect may always be found by subtraction of the preceding ones from zero. Secondly, the size of the effects simply reflects the size of the marginal totals; so that of the parameters $\hat{u}_{1(i)}$, $\hat{u}_{1(1)}$ is the largest

Table 5.3   *Estimated main effects for data in Table 5.1*

		Variable 1	Variable 2
	1	$\hat{u}_{1(1)} = 0.77$	$\hat{u}_{2(1)} = -0.50$
Category	2	$\hat{u}_{1(2)} = 0.07$	$\hat{u}_{2(2)} = 0.71$
	3	$\hat{u}_{1(3)} = -0.84$	$\hat{u}_{2(3)} = -0.21$

since the first category of variable 1 has the largest of the marginal totals of this variable. Similarly of the parameters $\hat{u}_{2(j)}$, $\hat{u}_{2(2)}$ is largest.

Setting $z_{ij} = \ln E_{ij}$ and adopting a 'bar' notation for means, that is

$$\bar{z}_{i.} = \frac{1}{c} \sum_{j=1}^{c} \ln E_{ij}, \text{etc.,}$$

the main effect estimates may be written in the form taken by parameter estimates in the analysis of variance:

$$\hat{u}_{1(i)} = \bar{z}_{i.} - \bar{z}_{..}, \qquad (5.20)$$
$$\hat{u}_{2(j)} = \bar{z}_{.j} - \bar{z}_{..}. \qquad (5.21)$$

Returning now to the three-dimensional data in Table 4.1 and the hypothesis that sex (variable 1) is independent of method (variable 2) and age (variable 3) jointly, considered in section 4.5. This hypothesis corresponds to the model

$$\ln F_{ijk} = u + u_{1(i)} + u_{2(j)} + u_{3(k)} + u_{23(jk)}. \qquad (5.22)$$

The expected values under the model are shown in Table 4.6 and from these estimates of the parameters in the model may be obtained. Adopting again the nomenclature introduced above, that is letting $z_{ijk} = \ln E_{ijk}$, etc., these estimates are given by

$$\hat{u} = \bar{z}_{...}, \qquad (5.23)$$
$$\hat{u}_{1(i)} = \bar{z}_{i..} - \bar{z}_{...}, \qquad (5.24)$$
$$\hat{u}_{2(j)} = \bar{z}_{.j.} - \bar{z}_{...}, \qquad (5.25)$$
$$\hat{u}_{3(k)} = \bar{z}_{..k} - \bar{z}_{...}, \qquad (5.26)$$
$$\hat{u}_{23(jk)} = \bar{z}_{.jk} - \bar{z}_{.j.} - \bar{z}_{..k} + \bar{z}_{...}. \qquad (5.27)$$

The estimated interaction parameters are shown in Table 5.4. Note that since $\hat{u}_{23(j.)} = 0$ and $\hat{u}_{23(.k)} = 0$, only 10 of the 18 interaction parameters have to be independently estimated, the remaining 8 being determined by these relationships.

Positive values of the parameters indicate positive associations between the corresponding categories of the table, negative values the reverse. For example, the parameter for the 'old' age category and hanging is 0.534; consequently more people will be found in this category than would be expected if the variables were independent. Of course, whether the difference is significant or simply attributable to random fluctuations will depend on whether or not the estimated parameter value is significantly different from zero; the question of the significance of parameter estimates will be discussed in section 5.6. The pattern of values in Table 5.4 reflects the differences,

Table 5.4 *Interaction parameter estimates for data in Table 4.1 under the partial independence model*

Method	Young	Age Middle	Old
Solid	0.113	− 0.018	0.131
Gas	0.605	0.084	− 0.688
Hang	0.595	0.061	0.534
Gun	0.081	0.057	− 0.138
Jump	0.305	− 0.217	0.522
Other	0.327	0.033	− 0.360

already seen in Chapter 4 (Table 4.6) between the observed frequencies and the expected values calculated under the hypothesis of the partial independence of the variables.

A difficulty which may arise when fitting log-linear models to contingency table data is the occurrence of zero cell entries. Since the logarithm of zero is minus infinity this causes obvious problems. Such entries may arise in two ways. The first is when it is impossible to observe values for certain combinations of variables, in which case they are known as *a priori* zeros and are discussed in Chapter 8. Secondly, they may arise owing to sampling variation when a relatively small sample is collected for a table having a large number of cells; in this case zero cell entries are known as sampling zeros. An obvious way to deal with the latter is to try to increase the sample size. When this is not possible, however, a commonly adopted procedure is to increase cell frequencies by the addition of a small constant, say 0.5, before proceeding with the analysis. In this way all zero entries are removed. Clearly this method would not be acceptable if there was a large number of sampling zeros since it would artificially increase the sample size. (Fienberg, 1969, discusses more formal methods for determining the size of the constant to be added to each cell frequency to remove sampling zeros.)

The analysis of multidimensional contingency tables using log-linear models often involves a large amount of computation, particularly where expected values have to be obtained iteratively (section 5.5). Consequently such analyses are invariably performed using some appropriate statistical package (see Appendix B).

5.4 Fixed marginal totals

In the preceding chapter it was mentioned that corresponding to particular hypotheses, particular sets of expected value marginal totals are constrained to be equal to the corresponding marginal totals of observed values. In terms of the parameters of the corresponding log-linear models this means that the u-terms included in the model determine the marginal constraints imposed on the expected values. For example, in a three-variable table, fitting a model including only main effects, that is:

$$\ln F_{ijk} = u + u_{1(i)} + u_{2(j)} + u_{3(k)}, \tag{5.28}$$

fixes the following

$$E_{i..} = n_{i..}, \quad E_{.j.} = n_{.j.}, \quad E_{..k} = n_{..k},$$

but no two-variable marginal totals are so constrained. In the case of the partial independence model, namely:

$$\ln F_{ijk} = u + u_{1(i)} + u_{2(j)} + u_{3(k)} + u_{23(jk)}, \tag{5.29}$$

the following equalities hold:

$$E_{i..} = n_{i..}, \quad E_{.j.} = n_{.j.}, \quad E_{..k} = n_{..k}, \quad E_{.jk} = n_{.jk}.$$

For some sets of data certain marginal totals of observed frequencies are fixed by the sampling design, so the corresponding u-term **must** be included in the model so that the corresponding marginals of expected values are similarly fixed. For example, Table 5.5 shows a data set previously considered by Bartlett (1935) and others, which gives the results of an experiment designed to investigate the propagation of plum root stocks from root cuttings. In this experiment the marginal totals $n_{.jk}$ are fixed *a priori* by the investigator at 240; consequently in any analysis of the data these marginals must be maintained at this value. Therefore when fitting log-linear models to these data, only models including the term u_{23} would be considered appropriate. (For more details see Bishop, 1969.)

Table 5.5   *Data on propagation of plum root stocks*

		Time of planting (variable 2)			
		At once		In Spring	
Length of cutting (variable 3)		Long	Short	Long	Short
	Alive	156	107	84	31
Condition of plant after experiment (variable 1)					
	Dead	84	133	156	209
		240	240	240	240

5.5 Obtaining expected values iteratively

As mentioned previously, expected values corresponding to some models cannot be obtained directly from particular marginal totals of observed values. (This is so because in such cases the maximum likelihood equations have no explicit solution.) Consequently, the expected values must be obtained in some other way. Bartlett (1935) was the first to describe a method of obtaining expected values for a model where they could not be calculated directly. More recently several authors, for example Bock (1972) and Haberman (1974), have suggested the Newton–Raphson method for this purpose. In this section, however, only the method of **iterative proportional fitting** originally given by Deming and Stephan (1940) will be considered. The method is described in detail in Bishop (1969). To illustrate the technique it will be applied to the data in Table 4.1 to obtain expected values under the model which specifies that there is no second-order iteration between the three variables, namely:

$$\ln F_{ijk} = u + u_{1(i)} + u_{2(j)} + u_{3(k)} + u_{12(ij)} + u_{13(ik)} + u_{23(jk)}. \quad (5.30)$$

For a three-dimensional table the model in (5.30) is the only one for which expected values are not directly obtainable.

Examining the terms in (5.30) shows that the totals $E_{ij.}$, $E_{i.k}$ and $E_{.jk}$ are constrained to equal the corresponding marginals of observed values. The iterative proportional fitting method begins by assuming a starting value, $E_{ijk}^{(0)}$ for each E_{ijk}, of unity and proceeds by adjusting these proportionally to satisfy the first marginal constraint $E_{ij.} = n_{ij.}$ by calculating

$$E_{ijk}^{(1)} = \frac{E_{ijk}^{(0)} \times n_{ij.}}{E_{ij.}^{(0)}}. \quad (5.31)$$

(Note that $E_{ij.}^{(1)} = n_{ij.}$.)

The revised expected values, $E_{ijk}^{(1)}$, are now adjusted to satisfy the second marginal constraint $E_{i.k} = n_{i.k}$ as follows:

$$E_{ijk}^{(2)} = \frac{E_{ijk}^{(1)} \times n_{i.k}}{E_{i.k}^{(1)}}. \quad (5.32)$$

(Note that $E_{i.k}^{(2)} = n_{i.k}$.)

The cycle is completed when the values given by (5.32) are adjusted to satisfy $E_{.jk} = n_{.jk}$ using

$$E_{ijk}^{(3)} = \frac{E_{ijk}^{(2)} \times n_{.jk}}{E_{.jk}^{(2)}}. \quad (5.33)$$

(Note that $E_{.jk}^{(3)} = n_{.jk}$.)

A new cycle now begins by using the values obtained from (5.33) in (5.31). The process is continued until differences between succeeding expected values differ by less than some very small value.

Applying the procedure to the values in Table 4.1 and remembering that it begins with values of unity in each cell, results in the following sequence of computations.

Cycle 1

Step 1. Using formula (5.31) gives

$$E^{(1)}_{111} = \frac{1 \times (398 + 399 + 93)}{(1 + 1 + 1)} = 296.67.$$

Similarly $E^{(1)}_{112} = 296.67$ and $E^{(1)}_{113} = 296.67$.

$$E^{(1)}_{121} = \frac{1 \times (121 + 82 + 6)}{(1 + 1 + 1)} = 69.67.$$

Similarly $E^{(1)}_{122} = 69.97$ and $E^{(1)}_{123} = 69.67$.

$$E^{(1)}_{211} = \frac{1 \times (259 + 450 + 154)}{(1 + 1 + 1)} = 287.67.$$

Similarly $E^{(1)}_{212} = 287.67$ and $E^{(1)}_{213} = 287.67$.

Step 2. Using formula (5.32) the procedure continues as follows:

$$E^{(2)}_{111} = \frac{296.67(398 + 121 + 455 + 155 + 55 + 124)}{(296.67 + 69.67 + 522.67 + 118.67 + 44 + 73.33)} = 343.43,$$

$$E^{(2)}_{112} = \frac{296.67(399 + 82 + 797 + 168 + 51 + 82)}{(296.67 + 69.67 + 522.67 + 118.67 + 44 + 73.33)} = 414.55,$$

$$E^{(2)}_{113} = \frac{296.67(93 + 6 + 316 + 33 + 26 + 14)}{(296.67 + 69.67 + 522.67 + 118.67 + 44 + 73.33)} = 128.12,$$

and so on. The estimates of the expected values after the first cycle and on convergence are given in Table 5.6.

This algorithm operates by proportionally fitting the marginal totals fixed by the model. When expected values can be obtained explicitly from appropriate marginal totals of observed values the fitting algorithm is clearly not essential, although it is easy to demonstrate that in such cases both methods give the same results.

Table 5.6 *Expected values under the hypotheses of no second-order interaction for data of Table 4.1*

Method			Young	Age Middle	Old
Male	Solid	1	383.4	399.4	107.2
		c	410.9	379.4	99.7
	Gas	1	128.7	72.5	7.7
		c	122.7	77.6	8.7
	Hang	1	447.3	817.5	303.2
		c	439.2	819.9	308.9
	Gun	1	169.4	156.8	29.8
		c	156.4	166.3	33.4
	Jump	1	52.0	53.9	26.1
		c	56.8	51.1	24.1
	Other	1	121.1	85.6	13.3
		c	122.0	84.7	13.3
Female	Solid	1	263.3	473.3	153.4
		c	246.1	469.6	147.3
	Gas	1	14.8	16.1	2.1
		c	13.3	17.4	2.3
	Hang	1	119.9	421.5	188.6
		c	110.8	427.1	192.1
	Gun	1	14.7	26.2	6.0
		c	12.6	27.7	6.6
	Jump	1	35.9	71.4	41.7
		c	38.2	70.9	39.9
	Other	1	41.3	56.1	10.6
		c	40.0	57.3	10.7

1: estimated expected value on iteration 1.
c: estimated expected value at convergence.

5.6 Numerical examples

5.6.1 *Coronary heart disease*

The data given in Table 5.7 concern coronary heart disease and have been discussed previously by Ku and Kullback (1974) and others. A total of 1330 patients were cross-classified with respect to the following three variables:

Variable	Level
1. Blood pressure	1. Less than 127 mm Hg
	2. 127–146
	3. 147–166
	4. > 167
2. Serum cholesterol	1. Less than 200 mg/100 cc
	2. 200–219
	3. 220–259
	4. > 260
3. Coronary heart disease	1. Yes
	2. No

The first model considered is that containing only main effects; such a model is equivalent to the hypothesis that the three variables are mutually independent so that the appropriate expected values may be calculated using (4.7). The likelihood ratio fit criterion, X_L^2, takes the value 78.96. The degrees of freedom associated with a

Table 5.7   *Coronary heart disease data*

	Blood pressure	Serum cholesterol				
		1	2	3	4	
Coronary heart disease	1	2	3	3	4	12
	2	3	2	1	3	9
	3	8	11	6	6	31
	4	7	12	11	11	41
		20	28	21	24	93
No coronary heart disease	1	117	121	47	22	307
	2	85	98	43	20	246
	3	119	209	68	43	439
	4	67	99	46	33	245
		388	527	204	118	1237
	Overall total	408	555	225	142	1330

particular model are found from

> d.f. = number of cells in table − number of parameters
> in fitted model that requires estimating.   (5.34)

In the case of a main effect model (see (5.28)) the following parameters must be estimated:

Parameters in model	Number of such parameters that require estimating
Overall mean effect, u	1
Main effect of variable 1 $u_{1(i)}$	$(r-1)$ (since $u_{1(.)} = 0$)
Main effect of variable 2 $u_{2(j)}$	$(c-1)$ (since $u_{2(.)} = 0$)
Main effect of variable 3 $u_{3(k)}$	$(l-1)$ (since $u_{3(.)} = 0$)

So a total of $r + c + l - 2$ parameters need to be estimated, and consequently from (5.34):

$$\text{d.f.} = rcl - r - c - l + 2.$$

For the data in Table 5.7, $r = 4$, $c = 4$ and $l = 2$, so the main effects model in this case has 24 d.f. Clearly a log-linear model including only main effect parameters does not provide an adequate fit for these data.

The next model considered was that given in (5.30), which specifies no second-order interaction between the three variables. In this case expected values have to be obtained iteratively using the algorithm described in the previous section. The model has an associated goodness-of-fit statistic $X_L^2 = 4.77$ with nine d.f.; since this is non-significant no second-order interaction need be postulated for these data. Table 5.8 shows the first-order interaction parameter estimates obtained for this model. In most cases when fitting models, interest focuses on the *simplest*, that is the one with fewest parameters that provides an adequate fit to the data. From the results derived above it is clear that a model somewhere between one involving only main effects and one involving all first-order interaction terms will be necessary for the coronary heart disease data. An indication of which of these alternative models might be most appropriate may be obtained by examining the **standardized values** in Table 5.8. These values are obtained by dividing a parameter estimate by its standard error. Details of how to compute the latter are given by Goodman

Table 5.8 *Some parameter estimates for the 'no second-order interaction' model fitted to the data of Table 5.7 (effects such as $\hat{u}_{13(12)}$ are obtained by using the fact that $\hat{u}_{13(1.)} = 0$)*

Variables		Magnitude of effect	Standard error	Standardized value
1 and 3	$\hat{u}_{13(11)}$	−0.219	0.111	−1.967*
	$\hat{u}_{13(21)}$	−0.238	0.106	−2.248*
	$\hat{u}_{13(31)}$	0.075	0.117	0.637
	$\hat{u}_{13(41)}$	0.383	0.120	3.186*
2 and 3	$\hat{u}_{23(11)}$	−0.227	0.125	−1.810
	$\hat{u}_{23(21)}$	−0.272	0.138	−1.970*
	$\hat{u}_{23(31)}$	0.054	0.095	0.576
	$\hat{u}_{23(41)}$	0.445	0.090	4.949*
1 and 2	$\hat{u}_{12(11)}$	0.222	0.205	1.083
	$\hat{u}_{12(12)}$	0.111	0.233	0.474
	$\hat{u}_{12(13)}$	−0.114	0.165	−0.687
	$\hat{u}_{12(14)}$	−0.219	0.159	−1.381
1 and 2	$\hat{u}_{12(21)}$	−0.180	0.202	−0.091
	$\hat{u}_{12(22)}$	−0.440	0.225	−0.193
	$\hat{u}_{12(23)}$	0.155	0.148	1.045
	$\hat{u}_{12(24)}$	−0.093	0.146	−0.638
1 and 2	$\hat{u}_{12(31)}$	−0.037	0.225	−0.163
	$\hat{u}_{12(32)}$	0.027	0.245	0.112
	$\hat{u}_{12(33)}$	−0.062	0.170	−0.364
	$\hat{u}_{12(34)}$	0.071	0.159	0.448
1 and 2	$\hat{u}_{12(41)}$	−0.167	0.234	−0.716
	$\hat{u}_{12(42)}$	−0.094	0.254	−0.372
	$\hat{u}_{12(43)}$	0.020	0.170	0.120
	$\hat{u}_{12(44)}$	0.241	0.159	1.518

*Indicates a 'significant' effect.

(1971), who also shows that the standardized values have, asymptotically, a standard normal distribution, and may therefore be compared with an appropriate normal deviate to obtain some idea as to the 'significance' of a particular effect. Usually comparing standardized values with ± 2 will be sufficient.

Examination of Table 5.8 suggests that a model including only the interaction terms u_{13} and u_{23} might provide an adequate fit for these data; consequently such a model, i.e.

$$\ln F_{ijk} = u + u_{1(i)} + u_{2(j)} + u_{3(k)} + u_{13(ik)} + u_{23(jk)} \qquad (5.35)$$

Table 5.9   *Some parameter estimates for the model specified in (5.35) fitted to the data of Table 5.7*

Variables		Magnitude	Standard error	Standardized value
1 and 3	$\hat{u}_{13(11)}$	− 0.262	0.117	− 2.233
	$\hat{u}_{13(21)}$	− 0.247	0.105	− 2.346
	$\hat{u}_{13(31)}$	0.084	0.117	0.720
	$\hat{u}_{13(41)}$	0.424	0.114	3.722
2 and 3	$\hat{u}_{23(11)}$	− 0.247	0.124	− 1.991
	$\hat{u}_{23(21)}$	− 0.281	0.138	− 2.029
	$\hat{u}_{23(31)}$	0.048	0.094	0.515
	$\hat{u}_{23(41)}$	0.480	0.090	5.333

was tried. The model in (5.35) is that of conditional independence discussed earlier; for these data it implies that there is no association between blood pressure and serum cholesterol for either coronary heart disease patients or those without coronary heart disease.

The expected frequencies for this model may be obtained explicity from

$$E_{ijk} = \frac{n_{i.k} n_{.jk}}{n_{..k}}. \tag{5.36}$$

Fitting the model leads to a value of X_L^2 of 24.40 with 18 d.f. which is not significant at the 5% level and so it seems that the conditional independence model in (5.35) does provide an adequate fit for these data. Some of the parameter estimates for the model are shown in Table 5.9. In general terms the results indicate that there is a positive association between high blood pressure (level 4) and the occurrence of coronary heart disease ($\hat{u}_{13(41)} = 0.424$, $p < 0.05$), and similarly a positive association between high serum cholesterol level and the presence of coronary heart disease ($\hat{u}_{23(41)} = 0.480$, $p < 0.05$). In addition, the lack of a second-order interaction implies that (a) the association between blood pressure and coronary heart disease is the same at all levels of serum cholesterol and (b) the association between serum cholesterol and heart disease is the same for all blood pressure levels.

5.6.2   Voting behaviour

For the next example of fitting log-linear models, the data given in Table 4.2 will be used. Here the data consists of a cross-classification

Table 5.10 *Models for the 'vote' data in Table 4.2*

Model	X^2	d.f.	p
[S] [C] [A] [V]	234.22	51	0.000
[SA] [C] [V]	233.43	47	0.000
[SV] [C] [A]	223.33	50	0.000
[S] [CV] [A]	82.82	49	0.002
[SA] [CV]	82.03	45	0.001
[SV] [CV] [AV]	46.80	44	0.358
[SV] [CVA]	22.92	28	0.737

of people by sex (S), class (C), age (A) and voting intention (V). Again, the first model considered was that of the mutual independence of the four variables forming the table. As discussed previously, such a model constrains the corresponding single variable marginal totals of expected and observed values to be equal, and so can be referred to in an obvious shorthand notation as [S], [C], [A], [V]. Similarly a model which postulated the absence of any four-variable or three-variable effects could be referred to as [SC], [SA], [SV], [CA], [CV], [AV], since such a model allows all two-variable effects, and the presence of these constrains the corresponding two-variable marginals of observed and expected values to be the same. A number of models were fitted to these data with the resulting goodness-of-fit statistics shown in Table 5.10.

There are several models which appear to provide an adequate fit and the problem of choosing which of these is the most appropriate is taken up in the next section.

5.7 Choosing a particular model

As the number of dimensions of a multidimensional table increases, so does the number of possible models, so that some procedures are clearly needed to indicate which models may prove reasonable for a data set and which are likely to be inadequate. One such procedure is to examine the standardized parameter values in the saturated log-linear model. These values may serve to indicate which parameters might be excluded and consequently which unsaturated models may be worth considering. In many cases, however, it will be found that several models provide an adequate fit to the data if judged simply by the significance of the usual goodness-of-fit statistics. In general the preferred model will then be the one with fewest parameters. In some cases, however, a test between rival models may be required to assess which gives the most parsimonious description of the data. In essence a way to choose between two competing goals is sought – a model complex enough to provide a good fit to the data and a model that is simple to interpret and 'smooths' rather than overfits the data.

Goodman (1971) and Fienberg (1970) show that for the hierarchical models considered in this chapter the difference in the X_L^2 values of two models may be used to assess the change in fit which results from the addition of extra parameters. For example, for the data in Table 5.7, the model specified in (5.35) and that specified by (5.30) both yield non-significant X_L^2 values. These two models differ by the presence of the u_{12} parameters in the latter. The model including these extra parameters would usually only be preferred over the simpler model if it provided a significantly *improved* fit. The difference in X_L^2 values in this case is 19.63 with nine d.f. (the difference in the degrees of freedom of the two models). The associated p value is 0.02 and it appears that the addition of the extra parameters provides a substantial improvement in fit and so the model which includes *all* first-order interaction terms is preferred for these data.

For model selection in more complex cases, Goodman (1971) proposed methods analogous to forward selection and backward elimination procedures used in multiple regression analysis. The forward selection process adds terms sequentially to the model until further additions do not improve the fit. At each stage the term giving the greatest improvement in fit is chosen. The backward elimination process begins with a more complex model and sequentially removes terms. At each stage the term removed is that causing the smallest decrease in the fit. The process stops when any further deletion leads to a significantly poorer-fitting model. Although both approaches can be useful, the following caveat issued by Agresti (1990) should be borne in mind. 'There is no guarantee that either strategy will lead to a meaningful model. Algorithmic selection procedures are no substitute for careful theory construction in guiding the formulation of models.'

To illustrate the forward and backward selection procedures the voting behaviour data will be used.

5.7.1 Forward selection

Beginning with the model corresponding to the mutual independence of the four variables, the procedure first considers the addition of each of the six possible first-order interaction terms. The term resulting in the largest reduction in the X_L^2 statistic [CV], the test statistic being reduced from 234.22 with 51 d.f. to 82.82 with 49 d.f. The remaining five first-order terms are now considered for adding to the current model; both [SV] and [AV] produce relatively large reductions in the test statistic and are retained to give at this stage the model [CV], [SV], [AV] which has $X_L^2 = 46.80$ with 44 d.f. The four second-order terms, [SCA], [SCV], [SAV] and [CAV], are now considered and the term for class/age/vote retained. The current model [CAV], [SV] has $X_L^2 = 22.92$ with 28 d.f. This is considered an adequate fit and the selection process is terminated.

The model indicates that there is a significant association between vote and sex and that the association between class and vote differs between age groups.

5.7.2 Backward selection

Taking the initial model as [SCV], [SCA], [SAV], [CAV], each second-order term is considered for removal. [SCV], [SCA] and [SAV] are all removed since they each give rise to only a small increase in the value of the test statistic. The procedure then begins to consider first-order terms for removal. Details are left as an exercise for the reader.

5.8 Displaying log-linear models graphically

The interpretation of a log-linear model fitted to a high-dimensional contingency table is often aided by means of a graphical display introduced by Darroch et al. (1980). In this display each variable is represented by a dot and those dots which are related through at least one non-zero interaction are connected by a line. If two variables are not connected at all in the diagram they are completely independent. If the connection between two variables can be broken by covering one or more variables, then the variables are conditionally independent given the covered variables. Figure 5.1, for example, shows a case of four variables, where variable 1 is independent of all other variables and variables 2 and 4 are independent given the value of variable 3.

A further illustration is provided by the analysis of a set of data obtained by questioning 600 individuals about six items relating to consumer behaviour reported by Andersen (1982). Various log-linear

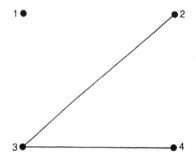

Fig. 5.1. *Graphical display for a four-dimensional contingency table.*

models were fitted to the resulting six-variable contingency table and an adequate fit was provided by a model in which all six-factor, all five-factor, all four-factor, all three-factor and a number of two-factor interactions were set to zero. The corresponding graphical display appears in Figure 5.2.

The interpretation of the model follows relatively simply from the diagram. For example, variable 2 is conditionally independent of variables 1 and 4 given variables 3 and 5. The same type of conditional

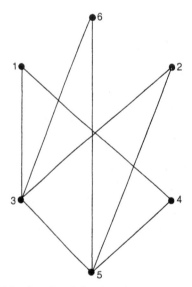

Fig. 5.2. *Graphical display for consumer data.*

independence applies to variable 6. Additionally, variable 2 is conditionally independent of variable 6 given variables 3 and 5. Variables 3 and 5 are clearly of major importance in this table.

5.9 Using correspondence analysis on multidimensional tables

In Chapter 3 correspondence analysis was introduced as a method for displaying graphically the residuals from the independence model for a two-dimensional contingency table. The method can be applied to a higher-way contingency table in several ways. The first would be simply to analyse separate two-dimensional tables for each category of other variables, so, for example, when a three-way table is of order $r \times c \times l$, l two-way tables could be subjected to correspondence analysis.

A more informative approach is to construct **interactive** variables by merging two or more of the original variables and apply correspondence analysis to the derived two-way table. For example, for a three-way table it is possible to consider three, two-way tables of order $r \times (c \times l)$, $c \times (r \times l)$ and $l \times (r \times c)$ respectively. Van der Hayden and de Leeuw (1985) demonstrate that correspondence analysis solutions from such tables graphically display **differences** between particular log-linear models. So, for example, correspondence analysis applied to the two-way table ($r \times l$) derived from a three-way table, can be interpreted in terms of the difference between log-linear models, [123] and [1] [23], that is the saturated model and one of partial independence.

Van der Heyden and de Leeuw also show that such correspondence analysis solutions do not show all the associations in the original table. For example, the solution obtained from the $r \times (c \times l)$ derived two-way table, says nothing about the interaction between variables 2 and 3; this interaction does not affect the solution. Consequently, when some first-order interaction is not thought to be particularly interesting, the corresponding two variables can be coded interactively.

To illustrate this use of correspondence analysis it will first be applied to the suicide data given in Table 4.1, combining age and sex as an interactive variable. The resulting two-way table to be analysed is shown in Table 5.11 and the two-dimensional correspondence analysis solution is shown in Figure 5.3. The first two axes account for 97.8% of the inertia and so clearly give an excellent summary of these data.

Table 5.11   *Age/sex by method of suicide*

| Age/sex | Method | | | | | |
	Solid	Gas	Hang	Gun	Jump	Other
Young/male (ym)	398	121	455	155	55	124
Young/female (yf)	259	15	95	14	40	38
Middle/male (mm)	399	82	797	168	51	82
Middle/female (mf)	450	13	450	26	71	60
Old/male (om)	93	6	316	33	26	14
Old/female (of)	154	5	185	7	38	10

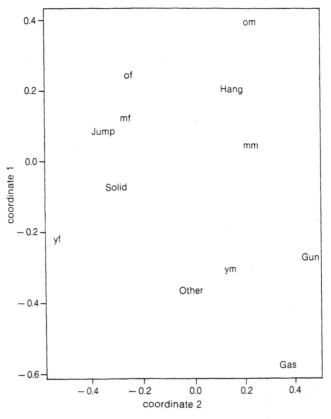

Fig. 5.3. *Suicide data: age/sex by method.*

Roughly, the first dimension stresses the differences in behaviour of men and women and the second the different use of methods by people of different ages. Both 'old' males and females use hanging as a method more often then expected, and young males, gun and gas. The result also implies that women tend to use jumping and solid as methods, men hanging, gas or guns.

5.10 Summary

In this chapter the fitting of log-linear models to contingency tables has been discussed at some length. Nevertheless, only a small part of the area has been covered and the chapter should be regarded as an introduction to more complete works such as Bishop *et al.* (1975) and Agresti (1990). The techniques that have been described allow a systematic approach to be taken to uncovering the relationships that may be present in complex multidimensional tables.

CHAPTER 6

Linear-logistic models

6.1 Introduction

Several authors, for example, Bhapkar and Koch (1968), have emphasized the difference between **factor** or **explanatory** variables which classify the unit of observation according to a description of the sub-population of the units to which it belongs (or to the experimental conditions which it undergoes), and **response** variables which classify according to a description of what happens to the unit during and/or after the experiment. The data in Table 5.5, for example, has essentially one response and two factor variables, the former being the state of the plant after the experiment, alive or dead, and the latter being time of planting and length of cutting. In any analysis of such data, the primary interest generally lies in assessing how the factor variables affect the response variable.

6.2 Models for data involving a response variable

To introduce the type of models which might be considered when a response variable is the essential feature of the data, the data shown in Table 6.1 will be used. These data arise from interest in the extent to which the recovery of patients is predictable from the symptoms they show when ill. To investigate the relationships, 819 male patients were assessed for the presence $(+)$ or absence $(-)$ of the three symptoms of depression, anxiety and delusions of guilt; at the end of a suitable time period each patient was then rated as having recovered or not from their illness.

For these data the effects of the explanatory variables on recovery could be investigated by looking at differences in the rates of recovery for different symptoms and symptom combinations. A simple method of making such an investigation would be to postulate possible models for the probability of recovery. For example, if **none** of the symptoms was suspected of affecting recovery a possible model for

Table 6.1 *Incidence of symptoms amongst psychiatric patients*

		Depression (variable 4)							
		−				+			
Anxiety (variable 3)		−		+		−		+	
Delusions (variable 2)		−	+	−	+	−	+	−	+
Condition of patient (variable 1)	recovered	68	3	58	3	70	23	129	59
	not recovered	137	3	70	3	69	10	87	27
Proportion recovered		0.3317	0.5000	0.4531	0.5000	0.5036	0.6970	0.5972	0.6860

the probabilities would be:

$$P^{234}_{jkl} = \theta, \tag{6.1}$$

where P^{234}_{jkl} is the probability of recovery for the cell arising from the jth category of variable 2 ($j = 1, 2$), the kth category of variable 3 ($k = 1, 2$) and the lth category of variable 4 ($l = 1, 2$). The observed proportions of patients recovering, p^{234}_{jkl}, given in Table 6.1, represent estimates of the probabilities. (In fact, $E(p^{234}_{jkl}) = P^{234}_{jkl}$.) The model in (6.1) specifies that the probability of recovery is the same for all cells in the table, in other words, for all combinations of the three symptoms. A more complicated (and more realistic) model arises if a parameter to represent the effect of, say, the symptom depression on recovery is introduced, to give

$$P^{234}_{jkl} = \theta + \theta_l. \tag{6.2}$$

Estimates of the parameters θ and θ_l ($l = 1, 2$) might now be obtained and the model tested for goodness of fit. A problem arises, however, from the necessity that the probabilities satisfy the condition:

$$0 \leqslant P^{234}_{jkl} \leqslant 1, \tag{6.3}$$

since parameter estimates could be obtained which lead to fitted or estimated probability values not satisfying (6.3). Because of this, and other problems (Cox and Hinkley, 1974), models for probabilities are not generally considered. A convenient method does, however, exist for representing the dependence of a probability on explanatory variables so that constraint (6.3) is inevitably satisfied, namely that of postulating models for the **logistic transformation** of the probability. This transformation is given by

$$\lambda_{jkl}^{234} = \ln[P_{jkl}^{234}/(1 - P_{jkl}^{234})]. \tag{6.4}$$

As the probability varies from 0 to 1, the corresponding logistic varies from $-\infty$ to $+\infty$.

Bishop (1969) and Goodman (1971) show how the log-linear models for frequencies discussed in the previous chapter may be adapted to fit linear models for the logistic function, i.e. **linear-logistic** models. For example, suppose a model is postulated which involves only a single 'overall mean parameter', that is:

$$\lambda_{jkl}^{234} = \theta. \tag{6.5}$$

Such a model is clearly equivalent to that specified in (6.1), although the parameters will not have the same values. In terms of the theoretical frequencies, F_{ijkl}, for Table 6.1:

$$P_{jkl}^{234} = F_{1jkl}/(F_{1jkl} + F_{2jkl})$$

$$= F_{1jkl}/F_{.jkl}. \tag{6.6}$$

The observed proportions of recovery are given by an equivalent relationship involving the observed frequencies, n_{ijkl}, namely:

$$p_{jkl}^{234} = n_{1jkl}/n_{.jkl}.$$

Substituting (6.6) in (6.4) gives

$$\lambda_{jkl}^{234} = \ln(F_{1jkl}/F_{2jkl}). \tag{6.7}$$

Consider now a log-linear model for the frequencies involving only an overall mean effect, and a main effect parameter for variable 1:

$$\ln F_{ijkl} = u + u_{1(i)}. \tag{6.8}$$

Substituting (6.8) in (6.7) gives

$$\lambda_{jkl}^{234} = u_{1(1)} - u_{1(2)} \tag{6.9}$$

which since $u_{1(1)} + u_{1(2)} = 0$ becomes

$$\lambda_{jkl}^{234} = 2u_{1(1)}, \tag{6.10}$$

which is equivalent to (6.5) with $\theta = 2u_{1(1)}$. So fitting the log-linear model specified in (6.8) is equivalent to fitting the model for the logistic function given in (6.5).

Now consider a more complicated logistic model, one in which the symptom depression is postulated to affect recovery:

$$\lambda_{jkl}^{234} = \theta + \theta_l. \tag{6.11}$$

An equivalent log-linear model is:

$$\ln F_{jkl} = u + u_{1(i)} + u_{4(l)} + u_{14(il)}, \tag{6.12}$$

since substituting this is (6.7) gives:

$$\lambda_{jkl}^{234} = [u_{1(1)} - u_{1(2)}] + [u_{14(1l)} - u_{14(2l)}], \tag{6.13}$$

which is seen to be of the same form as (6.11) with $\theta = [u_{1(1)} - u_{1(2)}]$ and $\theta_l = [u_{14(1l)} - u_{14(2l)}]$. The parameter u_{14} should now be regarded as quantifying the effect of the symptom depression on recovery. In a similar way other log-linear models can be shown to be equivalent to particular linear-logistic models. A complication arises when considering the estimation of parameters in such models. To illustrate the problem, consider adding to the model in (6.12) any parameter not involving variable 1, say u_2, to give:

$$\ln F_{ijk} = u + u_{1(i)} + u_{2(j)} + u_{4(l)} + u_{14(il)}. \tag{6.14}$$

Substituting this in (6.7) again gives (6.13), so that (6.14) is also equivalent to the logistic model defined in (6.11). Inclusion of any other u-term not involving variable 1 in the model for log-frequencies will also lead to (6.13); each model will, however, lead to different estimates of u_1 and u_{14} and consequently of θ and θ_l; what is required is to choose the log-linear model that is equivalent to the required logistic model and gives identical parameter estimates to those that would be obtained from fitting such models directly (see the next section). Bishop (1969) shows that the appropriate log-linear model is that which includes the term $u_{IJKL...(ijkl...)}$ where I, J, K, L, etc., represent factor variables. So the log-linear model required to produce the logistic model in (6.11) is:

$$\ln F_{ijkl} = u + u_{1(i)} + u_{14(il)} + u_{234(jkl)}$$
$$+ \text{all other implied lower-order terms}, \tag{6.15}$$

and that for the logistic model in (6.5) is

$$\ln F_{ijkl} = u + u_{1(i)} + u_{234(jkl)}$$
$$+ \text{all other implied lower-order terms}. \tag{6.16}$$

Fitting the second of these to the data in Table 6.1 by the methods described in Chapter 5 leads to a value of X_L^2 of 50.44 with seven d.f. Clearly this model does not fit adequately. Fitting (6.15) gives

$X_L^2 = 14.87$ with six d.f., a substantially improved fit; the estimate of the parameter $u_{14(11)}$ is -0.214 with standard error 0.081. In clinical terms this implies that prognosis is good for patients who originally have the symptom depression present. If an improved fit was required, additional parameters could be fitted in the same way. Terms such as u_{123}, if needed in the model to provide an adequate fit, would now be indicative that variables 2 and 3 were not acting independently on the response variable (variable 1).

6.3 Fitting linear-logistic models directly

The introduction to linear-logistic models given in the previous section shows how they can be fitted to data through the use of equivalent log-linear models. It is, however, possible to fit such models directly, most commonly by maximum likelihood methods. In addition it is unnecessary to restrict the models to data sets for which the explanatory variables are categorical, so that a linear-logistic model can be thought of more generally as a suitable **regression** model for a dichotomous response variable; viewed in this way the model may reformulated in terms of a response variable y taking values 0 or 1, and a set of explanatory variables, $x_1, x_2, \ldots, x_p$, as:

$$E(y) = Pr(y = 1) = \frac{\exp[\beta_0 + \Sigma_{i=1}^p \beta_i x_i]}{1 + \exp[\beta_0 + \Sigma_{i=1}^p \beta_i x_i]}, \qquad (6.17)$$

leading to the following linear model for the logistic function

$$\ln \frac{Pr(y = 1)}{Pr(y = 0)} = \beta_0 + \sum_{i=1}^p \beta_i x_i, \qquad (6.18)$$

where $\beta_0, \beta_1, \cdots \beta_p$ are regression coefficients to be estimated.

To illustrate the use of logistic regression when the explanatory variables include both categorical and interval scaled variables, the data shown in Table 6.2 will be used. Here the response variable consists of the diagnostic group to which a patient belongs, dementia or depression. The explanatory variables are sex, education and age. Education has four categories labelled 1, 2, 3 and 4. Clearly, it is not sensible to include these directly in any analysis since the implication would be that a change from say level 1 to 2 has the same effect as a change from 3 to 4. Consequently, the categories are recoded prior to analysis as three 'dummy' variables as follows:

	Dummy variable		
	D_1	D_2	D_3
Education			
1	0	0	0
2	1	0	0
3	0	1	0
4	0	0	1

Estimated regression coefficients for these dummy variables will now indicate the effect of each educational level compared to level 1. The estimated coefficients are given in Table 6.3 and the model is as follows:

$$\ln \frac{Pr(\text{dementia})}{Pr(\text{depression})} = 5.89 + 0.0779\,\text{age} - 0.2606\,\text{sex}$$

$$+ 1.408\,D_1 - 0.1285\,D_2 - 0.2791\,D_3, \quad (6.19)$$

Table 6.2 *Dementia and depression data*

Group	Age	Sex $(1 = M, 2 = F)$	Education
Depressi	77	2	3
Depressi	73	1	3
Dementia	87	2	3
Depressi	75	1	1
Dementia	76	1	4
Dementia	78	2	3
Depressi	73	2	4
Dementia	71	2	3
Dementia	87	2	4
Dementia	77	2	3
Dementia	81	2	2
Dementia	85	1	3
Dementia	62	2	1
Dementia	73	2	3
Dementia	88	2	4
Depressi	76	2	4
Dementia	85	2	3
Depressi	81	2	3

(Contd.)

Table 6.2 (*Contd.*)

Group	Age	Sex (1 = M, 2 = F)	Education
Depressi	67	2	4
Depressi	80	2	4
Depressi	77	1	4
Depressi	74	2	1
Dementia	73	2	4
Dementia	76	2	2
Dementia	82	1	3
Dementia	84	2	3
Depressi	72	1	3
Dementia	66	1	2
Dementia	78	1	4
Depressi	64	2	3
Dementia	85	1	3
Dementia	68	2	3
Dementia	86	2	3
Dementia	75	1	4
Dementia	76	1	3
Depressi	85	1	3
Depressi	63	2	4
Depressi	82	2	4
Depressi	65	1	3
Dementia	61	1	1
Depressi	77	2	3
Depressi	83	1	3
Depressi	68	2	3
Dementia	76	1	3
Depressi	70	2	3
Depressi	69	2	3
Depressi	76	2	4
Dementia	82	1	3
Dementia	68	1	1
Depressi	61	2	3
Depressi	74	2	3
Dementia	83	1	2
Depressi	76	1	2
Depressi	67	2	3
Dementia	83	2	4
Depressi	71	1	4
Depressi	83	1	3
Depressi	89	2	3
Depressi	90	1	3
Depressi	68	1	1
Dementia	87	1	3

(*Contd.*)

Table 6.2 (*Contd.*)

Group	Age	Sex (1 = M, 2 = F)	Education
Depressi	81	2	1
Depressi	74	2	3
Depressi	81	2	3
Dementia	69	2	3
Depressi	86	2	3
Dementia	75	1	4
Dementia	75	1	3
Depressi	85	1	3
Depressi	63	2	4
Depressi	82	2	4
Depressi	65	1	3
Dementia	61	1	1
Depressi	77	2	3
Depressi	83	1	3
Depressi	68	2	3
Dementia	76	1	3
Dementia	65	1	3
Depressi	70	2	3
Depressi	69	2	3
Depressi	76	2	4
Dementia	82	1	3
Dementia	68	1	1
Depressi	61	2	3
Depressi	74	2	3
Dementia	83	1	2
Depressi	76	1	2
Depressi	67	2	3
Dementia	83	2	4
Depressi	71	1	4
Depressi	83	1	3
Depressi	89	2	3
Depressi	90	1	3
Depressi	68	1	1
Dementia	87	1	3
Depressi	81	2	1
Depressi	74	2	3
Depressi	81	2	3
Dementia	69	2	3
Depressi	77	1	3
Depressi	72	2	3
Dementia	83	2	3
Dementia	87	2	4
Dementia	78	1	1

(*Contd.*)

Table 6.2   (Contd.)

Group	Age	Sex (1 = M, 2 = F)	Education
Depressi	78	2	1
Depressi	87	2	3
Depressi	79	1	4
Depressi	74	2	3
Depressi	80	1	3
Depressi	80	2	3
Depressi	76	1	1
Depressi	69	2	4
Dementia	65	1	1
Dementia	82	2	3
Depressi	68	2	4
Dementia	88	2	3
Depressi	71	2	4
Depressi	74	2	3
Depressi	83	1	4
Dementia	85	2	3
Dementia	88	1	3
Dementia	77	2	1
Depressi	72	1	1
Dementia	73	1	3
Depressi	79	2	3
Depressi	67	2	1
Dementia	71	2	3
Dementia	80	2	1
Dementia	78	1	3
Dementia	84	2	1
Depressi	68	2	3
Depressi	70	1	3
Dementia	81	1	3
Dementia	78	2	3
Dementia	75	1	3
Dementia	78	1	3
Dementia	82	1	2
Dementia	82	1	3
Depressi	74	2	1
Depressi	77	2	3
Dementia	86	2	4
Dementia	85	2	4
Dementia	69	1	3
Depressi	74	1	3
Dementia	84	1	2
Dementia	85	2	3
Dementia	88	1	3

(Contd.)

Table 6.2 (*Contd.*)

Group	Age	Sex (1 = M, 2 = F)	Education
Dementia	77	2	1
Depressi	72	1	1
Dementia	73	1	3
Depressi	79	2	3
Depressi	67	2	1
Dementia	71	2	3
Dementia	80	2	1
Dementia	78	1	3
Dementia	84	2	1
Depressi	68	2	3
Depressi	70	1	3
Dementia	81	1	3
Dementia	78	2	3
Dementia	75	1	3
Dementia	78	1	3
Dementia	82	1	2
Dementia	82	1	3
Depressi	74	2	1
Depressi	77	2	3
Dementia	86	2	4
Dementia	85	2	4
Dementia	69	1	3
Depressi	74	1	3
Dementia	84	1	2
Depressi	77	1	3
Dementia	80	2	3
Depressi	78	2	3
Dementia	84	2	3
Dementia	82	2	3
Dementia	85	2	3
Depressi	83	2	3
Depressi	81	2	3
Depressi	77	2	4
Dementia	83	1	3
Dementia	74	2	3
Dementia	81	1	2
Depressi	79	1	3
Dementia	81	2	4
Depressi	66	2	3
Dementia	78	2	3
Dementia	68	1	2
Dementia	73	2	4
Dementia	72	2	4
Depressi	78	1	1

where D_1, D_2 and D_3 are the dummy variables coding educational level.

Comparing each of the coefficients in Table 6.3 to their standard errors suggests that sex and education are not needed in the model. Refitting the model considering only age as an explanatory variable gives

$$\ln \frac{Pr(\text{dementia})}{Pr(\text{depression})} = -6.19 + 0.079 \, \text{age}. \tag{6.20}$$

Exponentiating the regression coefficient for age gives the value 1.08. The model may now be interpreted as suggesting that each increase in age of one year gives an 8% increase in the probability of a patient being diagnosed as demented rather than depressed. The 95% confidence interval for the exponentiated coefficient is 1.02 to 1.15, so this increased probability lies between 2% and 15%.

When there are many explanatory variables the problem of choosing the 'best' model from the many possibilities arises (cf. section 5.7). An interesting graphical procedure which may be helpful is suggested by Fowlkes et al. (1988). The main feature of the method is a scatterplot of values of the likelihood ratio goodness-of-fit criterion for each model against the corresponding degrees of freedom. Each point on this (d.f., X_L^2) plot represents a fitted model. Two lines are superimposed on the model; the first for points (d.f., d.f.) and the second for points (d.f. $\chi^2_{0.95 \, \text{d.f.}}$). Plotted points that fall near the lower line (d.f., d.f.) represent models that fit the data well, since $E(X_L^2) = \text{d.f.}$ if the model is correct and the sample size is large. Points that fall above the upper line (d.f., $\chi^2_{0.95, \text{d.f.}}$) represent models that are significantly poor fits. Points to the right of the plot (large d.f., relatively few parameters) represent simple models. The main aim of the plot is to display graphically the trade-off between model

Table 6.3 *Parameter estimates and standard errors for logistic regression model fitted to data in Table 6.2*

Term	Coefficient	Standard error	Coefficient/SE
Age	0.078	0.033	2.330
Sex	−0.261	0.385	0.680
Education			
D_1	1.408	1.290	1.090
D_2	−0.128	0.679	−0.189
D_3	−0.279	0.786	−0.035

Table 6.4 *Women and mathematics data: number of high-school students responding to the statement 'I'll need mathematics in my future work'*

| | Suburban school | | | | Urban school | | | |
| | Female | | Male | | Female | | Male | |
	Lecture	No lecture	Lecture	No lecture	Lecture	No lecture	Lecture	No lecture
Plans: college; preference: math–science								
Agree	37	27	51	48	51	55	109	86
Disagree	16	11	10	19	24	28	21	25
Plans: college; preference: liberal arts								
Agree	16	15	7	6	32	34	30	31
Disagree	12	24	13	7	55	39	26	19
Plans: job; preference: math–science								
Agree	10	8	12	15	2	1	9	5
Disagree	9	4	8	9	8	9	4	5
Plans: job; preference: liberal arts								
Agree	7	10	7	3	5	2	1	3
Disagree	8	4	6	4	10	9	3	6

Reproduced from Fowlkes *et al.* (1988) by permission of the American Statistical Association.

simplicity and goodness-of-fit. To illustrate the technique the data in Table 6.4 (taken with permission from Fowlkes *et al.*, 1988) will be used. The plot for 19 fitted models is shown in Figure 6.1. (In general, of course, many more models would be tried but the few fitted here serve to demonstrate the procedure). Examination of the plot shows that the points representing models M 16, M 12 and M 11 fall just below the (d.f. $\chi^2_{0.95\,d.f.}$) line and are also to the right of the plot. M 12 includes the sex, teach and plans variables, M 11 the variables teach and plans and their interaction and M 16 the variables sex and plans and their interaction. It is these models which would now be investigated in more detail. This is left as an exercise for the reader.

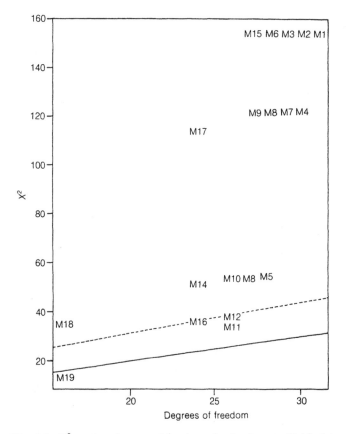

Fig. 6.1. X^2 *against degrees of freedom plot for data in Table 6.4.*

6.4 Polychotomous logistic regression

Although linear-logistic models are most commonly used in situations involving a dichotomous response variable, there are occasions when a similar model is required for data involving a categorical response with more than two categories. For example, Table 6.5 shows a further set of data concerned with the voting behaviour of individuals; in this case, however, the number of liberal-democrat supporters was also observed. Interest lies in investigating how the explanatory variables age, class and sex influence the three category response variable, party voted for.

The revised linear-logistic model for this situation has the form

$$Pr(y=j) = \frac{\exp v_j}{1 + \Sigma_{i=1}^{J-1} \exp v_j}, \tag{6.21}$$

for $j = 1,2,\cdots,(J-1)$ and

$$Pr(y=J) = 1 - \sum_{j=1}^{J-1} Pr(y=j), \tag{6.22}$$

where J is the number of categories of the response variable and v_j is a linear function of the explanatory variables (which again can be a mixture of categorical and continuous), given by:

$$v_j = \beta_{0j} + \sum_{i=1}^{p} \beta_{ij} x_i. \tag{6.23}$$

As before the parameters in the model (the regression coefficients), can be estimated by maximum likelihood methods as described in Green (1984), Koch and Edwards (1985) and McCullagh (1980). Fitting the model to the data in Table 6.5 and using a stepwise procedure (Chapter 5) to select which terms should be included gives the results shown in Table 6.6. Age, class and sex each have a significant effect on voting behaviour. (Both class and age have been recoded in terms of two dummy variables.)

Interpreting the coefficients in such complex models is not always straightforward; here, however, the explanation is relatively clear. For example, the negative coefficient of sex (coded 0 for males and 1 for females), for both the conservative and labour categories, indicates that a change from sex = 0 (i.e. male) to sex = 1 (i.e. female) decreases the relative probability of being a Conservative or Labour supporter in favour of supporting the Liberal-Democrats. This decrease is very small and not significant amongst conservative supporters, but is substantial amongst labour voters.

Table 6.5 *Voting behaviour*

	Male			Female		
	Conservative	Labour	Liberal-Democrat	Conservative	Labour	Liberal-Democrat
Upper middle class						
Age group						
A1	31	8	3	36	9	4
A2	44	16	8	53	18	10
A3	7	6	4	7	3	5
Lower middle class						
Age group						
A1	29	17	2	42	10	5
A2	41	27	6	62	12	10
A3	9	9	4	13	7	12
Working class						
Age group						
A1	43	77	2	69	57	6
A2	61	141	8	68	137	10
A3	14	34	1	18	33	3

Table 6.6 *Parameter estimates for logistic regression applied to data in Table 6.4*

		Coefficient	Standard error	Coefficient/SE	Exp. (coeff.)	95% confidence interval Low	High
Outcome: Conservative							
Age	(1)	−0.58	0.27	−2.16	0.56	0.33	0.95
	(2)	−0.156	0.32	−4.94	0.21	0.11	0.39
Class	(1)	0.03	0.26	0.13	1.03	0.62	1.71
	(2)	0.57	0.27	2.10	1.77	1.04	3.0
Sex		−0.27	−0.22	−1.21	0.76	0.49	1.17
Const. 1		2.37	0.30	7.83	10.70	5.94	19.26
Outcome: Labour							
Age	(1)	−0.16	0.28	−0.56	0.86	0.50	1.47
	(2)	−0.85	0.32	−2.62	0.43	0.23	0.81
Class	(1)	0.25	0.29	0.84	1.28	0.72	2.27
	(2)	2.24	0.29	7.77	9.39	5.32	16.58
Sex		−0.74	0.23	−3.23	0.48	0.31	0.75
Const. 2		1.20	0.32	3.69	3.32	1.77	6.22

Moving on now to the coefficients for class, particularly that comparing working class to upper middle, indicates that the relative probabilities of being a Labour voter rather than a Liberal-Democratic voter, is far more in favour of Labour in the working class, than amongst upper middle class voters.

Further interpretation is left as an exercise for the reader.

6.5   Logistic models for case-control studies

As mentioned in Chapter 2 one-to-one matching is frequently used by research workers to increase the precision of a comparison. For comparing such matched samples on a single variable, McNemar's test as described in Chapter 2 is appropriate. For the matched pair case-control study in which p risk variables are under investigation, Woolson and Lachenbruch (1982) suggest a method of analysis based on the linear-logistic model described previously in this chapter. The data from such a study takes the general form shown in Table 6.7. For a person with risk factor values $x_{i1}, x_{12}, ..., x_{ip}$ the probability of the condition of interest is modelled by the linear-logistic function given in (6.17), and regression coefficients which indicate differences between cases and controls may be estimated by a **conditional likelihood** procedure. Essentially the approach involves the same estimation process as logistic regression but here the 'response variable' always takes the value of one. Although technical details are outside the scope of this text, the procedure may be illustrated by means of an example. For this purpose, the data shown in Table 6.8 will be used. These data arise from a study designed to investigate whether known risk factors for the development of schizophrenia may be more common in schizophrenics with a low

Table 6.7   *Notation for matched pair case-control data with p risk variables*

Pair	Case	Control
1	$(x_{11}, ..., x_{1p})$	$(y_{11}, ..., y_{1p})$
2	$(x_{21}, ..., x_{2p})$	$(y_{21}, ..., y_{2p})$
⋮	⋮	⋮
i	$(x_{i1}, ..., x_{ip})$	$(y_{i1}, ..., y_{ip})$
⋮		
n	$(x_{n1}, ..., x_{np})$	$(y_{n1}, ..., y_{np})$

Table 6.8   *Case-control study: risk factors for schizo-phrenia*

Pair	Case 1	Case 2	Control 1	Control 2
1	0	1	0	0
2	1	0	1	0
3	0	0	0	0
4	1	1	1	0
5	0	0	1	0
6	0	0	0	0
7	0	0	0	0
8	1	0	1	0
9	0	1	0	0
10	0	0	0	0
11	0	0	1	0
12	1	0	1	0
13	0	0	1	0
14	0	0	0	0
15	0	1	0	0
16	0	0	0	0
17	1	0	0	0
18	0	0	0	1
19	0	0	0	1
20	0	0	1	0
21	1	1	0	0
22	0	0	0	0
23	0	0	0	1
24	1	1	0	0
25	0	1	0	0
26	0	0	1	1
27	0	1	1	0
28	1	0	0	0
29	0	0	0	0
30	1	0	0	0
31	0	0	0	0
32	0	1	0	0
33	1	0	0	0
34	0	0	0	0
35	1	0	0	0
36	0	0	0	0

1 = Birth complication, 2 = History.

Table 6.9 *Parameter estimates and standard errors for logistic model on case-control data*

Variable	Coefficient	Standard error	Coefficient/SE
1. Birth Complication	0.805	0.607	1.326
2. History	0.038	0.574	0.066

age of onset compared to those with age of onset in the commonly occurring range. The risk factors considered were complications during pregnancy and/or birth and family history of psychosis, both rate simply as 'present' (1) or 'absent' (0). Thirty-six people with a strict adult diagnosis of schizophrenia and who had been seen for any reason by a child psychiatrist before the age of 16 were identified as the low age of onset cases. Controls who also had schizophrenia, but who had not been seen by the psychiatric service until after the age of 21 were matched one-to-one with the cases on sex, race and socio-economic class. The estimated coefficients for the logistic model and their standard errors are shown in Table 6.9. For these data neither of the coefficients are significantly different from zero and so there is no evidence that patients with early onset of schizophrenia differ from those with late onset with respect to the two risk factors considered.

6.6 Summary

Linear-logistic models are extremely useful in investigating the effects of explanatory variables on a categorical response. Most common are situations involving a two category response variable, but appropriate models are now also available when the response variable has more than two categories. A detailed account of such models is available in Hosmer and Lemeshow (1989).

CHAPTER 7

Contingency tables with ordered categories

7.1 Introduction

Contingency tables formed by variables having classification categories which fall into a natural order, for example, severity of disease, age group, amount of smoking, etc., may be regarded as frequency tables for a sample from a bivariate population where the scales for the two underlying continuous variables have been divided into r and c categories respectively. Making such an assumption often allows the variables to be quantified by allotting numerical values to the categories and subsequently the use of regression techniques to detect linear and higher-order trends in the table. In this way **specific** types of departure from independence may be considered, leading to more sensitive tests than those obtained by the usual chi-square statistic. The assigning of arbitrary scores to categories, followed by the use of regression techniques is perhaps the simplest way of dealing with contingency tables with ordered categories. The method is described in section 7.2. Alternatives to the assignment of scores are, however, available which are particularly useful when the table has more than two variables. Many of these have been developed over the last decade or so and involve adaptations of log-linear or linear-logistic models. Such methods are described in sections 7.3 and 7.4.

7.2 The analysis of ordered tables by assigning scores to categories

To illustrate this approach to the analysis of tables with ordered categories, the data in Table 7.1 will be used. These data consist of a sample of 223 boys classified according to age and to whether or not they had been rated as inveterate liars. An overall chi-square test on Table 7.1 gives a value of 6.69 with four d.f. ($p = 0.15$), leading to the conclusion that there is no association between age and lying.

This is, however, a test which covers all forms of departure from independence and so is relatively insensitive to departures of a specified form. Here the examination of the proportion of inveterate liars in each age group, namely:

0.286   0.367   0.380   0.458   0.568

suggests that the proportion increases steadily with age; consequently, a test specifically designed to detect a trend in these proportions is likely to be more sensitive.

To arrive at such a test, numerical values are first assigned to the classification categories. This has been done in Table 7.1 where the age groups have been allocated scores running from -2 to $+2$, and the lie scale has been quantified by allotting the value $+1$ to the category 'inveterate liar' and the value 0 to 'non-liars'. These quantitative values are chosen quite arbitrarily. They are evenly spaced in the present example but they need not always be so. For example, if it was thought that lying was especially associated with puberty and the immediate post-pubertal period, the scores for age might be taken as $-2, -1, 0, 3, 6$ or some other values which give greater weight to the age groups in which there was special interest. The reason for choosing -2 as the first score has no significance other than it helps to keep the arithmetic simple since the five scores -2 to 2 sum to 0.

Having quantified the data, they are now treated as arising from a bivariate frequency table and estimates of correlation coefficients or regression coefficients obtained. For example, to find the linear regression coefficient of lying (y) on age (x), the usual formula is used giving:

$$b_{yx} = \frac{C_{yx}}{C_{xx}}, \qquad (7.1)$$

Table 7.1   *Boys' ratings on a lie scale*

| | | Age group | | | | |
| | | 5–7 | 8–9 | 10–11 | 12–13 | 14–15 |
Score:		-2	-1	0	1	2
Inveterate liars	1	6	18	19	27	25	95
Non-liars	0	15	31	31	32	19	128
		21	49	50	59	44	223

where

$$C_{yx} = \sum_{i=1}^{2} \sum_{j=1}^{5} n_{ij} y_i x_j - \left(\sum_{i=1}^{2} n_{i.} y_i \right) \left(\sum_{j=1}^{5} n_{.j} x_j \right) / N,$$

$$C_{xx} = \sum_{j=1}^{5} n_{.j} x_j^2 - \left(\sum_{j=1}^{5} n_{.j} x_j \right)^2 / N,$$

$$C_{yy} = \sum_{i=1}^{2} n_{i.} y_i^2 - \left(\sum_{i=1}^{2} n_{i.} y_i \right)^2 / N.$$

The variance of the regression coefficient is given by:

$$V(b_{yx}) = \frac{C_{yy}}{N C_{xx}}. \tag{7.2}$$

For the data in Table 7.1 the values are $C_{yx} = 23.14$, $C_{xx} = 353.94$, $C_{yy} = 54.53$, leading to

$$b_{yx} = 0.065\,38,$$
$$V(b_{yx}) = 0.000\,690\,9.$$

The component of chi-square due to linear due to linear trend is given by $b_{yx}^2/V(b_{yx})$, that is 6.1887. This has a single degree of freedom; recalling that the overall chi-square value for these data is 6.691 based on four degrees of freedom; the following table may now be drawn up:

Source of variation	d.f.	X^2
Linear regression of lying on age	1	6.189 $p < 0.025$
Departure from regression		
(by subtraction)	3	0.502 non-significant
		6.691

So although the overall X^2 statistic with four degrees of freedom is not significant, the X^2 value due to regression, based on a single degree of freedom, is significant beyond the 2.5% level. Partitioning the overall value in this way has greatly increased the sensitivity of the test; returning to the data in Table 7.1 the conclusion to be drawn is that there is a significant increase in lying with increase in age for the age range in question. Clearly this increase is linear rather than

Table 7.2

| Malformation | Maternal alcohol consumption (drinks/day) | | | | | |
	0	< 1	1–2	3–5	≥ 6	
Absent	17 066	14 464	788	126	37	32 481
Present	48	38	5	1	1	93
	17 114	14 502	793	127	38	32 574

Taken with the permission of the Biometrics Society from Graubard and Korn (1987).

curvilinear since departure from linear regression is represented by a chi-square value of only 0.502 with three degrees of freedom, which is a long way from being significant. (An interesting fact, pointed out by Yates, is that the partition of chi-square obtained above is the same whether the regression coefficient of y and x or that of x on y is used.)

The scores assigned to categories in an ordered table can, of course, considerably affect the inferences drawn. For example, data from a recent study of the effect of maternal alcohol consumption on congenital sex organ malformations (Graubard and Korn, 1987), are given in Table 7.2. If equally spaced scores, 1, 2, 3, 4 and 5, are used, the p value for the test of linear trend takes the value 0.1765. If, however, the average number of drinks/day, 0, 0.5, 1.5, 4 and 8, are used as the column scores, the p value becomes 0.0078. Consequently, the 'average drinks per day' scores, unlike the 'equally spaced' scores, reveal that the risk of malformations increases significantly with increasing alcohol consumption.

The discussion above has been in terms of testing for trend in a $2 \times c$ table. The method described is, however, also applicable to the general contingency table with more than two rows, when both classifications are ordered. Bhapkar (1968) gives alternative methods for testing for trends in contingency tables, which in most cases will differ little from the method discussed above. An investigation of the power of chi-square tests for linear trends has been made by Chapman and Nam (1968).

7.3 Log-linear models for tables with ordered categories

The log-linear models discussed in Chapter 5 treated all classifications as nominal, in the sense that parameter estimates and chi-square statistics are **invariant** to orderings of categories. Consequently the models as thus formulated ignore important information when at least one variable is ordinal. The models are, however, easily adapted to deal with contingency tables containing ordered variables by replacing the previously used 'deviations from overall mean' u-parameters, by parameters representing linear, quadratic and, if appropriate, higher order effects. The process is greatly simplified if the levels of the ordered variables can be assumed to be equally spaced, in which case **orthogonal polynomials** may be used. For example, for an ordered variable having three equally spaced categories, the linear and quadratic effects are obtained from the usual u-parameters by applying the orthogonal polynomial coefficients to be found, for example, in Kirk (1968, Table D.12), to give the following:

$$\text{Linear main effect, } u_{1(L)} = \tfrac{1}{2}[(1)u_{1(1)} + (0)u_{1(2)} + (-1)u_{1(3)}]$$

$$= \tfrac{1}{2}[u_{1(1)} - u_{1(3)}] \tag{7.3}$$

assuming that the first level of the variable is the 'high' level.

$$\text{Quadratic main effect, } u_{1(Q)} = \tfrac{1}{4}[(1)u_{1(1)} - (2)u_{1(2)} + (1)u_{1(3)}]$$

$$= \tfrac{1}{4}[u_{1(1)} + u_{1(3)} - 2u_{1(2)}]. \tag{7.4}$$

These parameters will represent trends in the single variable marginal totals of the ordered variable. Similarly, linear and quadratic effects may be found for the interaction between the ordered and the other unordered variables; the size of such effects indicates the similarity

Table 7.3 *Social class and number of years in present occupation at first attendance as out-patients for a sample of patients diagnosed as either neurotics or schizophrenics*

| | Diagnosis (variable 2) | | | | | | |
	Neurotic			Schizophrenic			
Duration of present occupation, years (variable 1)	<2	2–5	5+	<2	2–5	5+	
Social class (variable 3) I + II	4	6	18	5	10	8	
III	12	17	37	13	18	23	
IV + V	8	5	3	12	7	5	
	24	28	58	30	35	36	211

or otherwise of the trend of the ordered variable in different categories of the unordered variable. To help to clarify this approach, the data shown in Table 7.3 will be used. These data, collected during an investigation into the environmental courses of mental disorder, show the social class and number of years in present occupation for a sample of patients diagnosed either as neurotic or schizophrenic.

Variable 1 (duration of present occupation) falls into a natural order and it may be informative to investigate possible trends in this variable using log-linear models with linear and quadratic effect parameters. Fitting a model specifying zero second-order interaction results in the parameter estimates shown in Table 7.4 and a value of X_L^2 of 2.00 which with four d.f. is non-significant. Examining first the main effects for variable 1, only that representing linear trend approaches significance; the quadratic effect is very small. These two effects reflect the trend in the overall number of patients in the three categories of the ordered variable that is:

Duration of present occupation	< 2 yr	2–5 yr	5 + yr
Number of patients	54	63	94

Table 7.4 *Parameter estimates obtained from fitting a log-linear model specifying zero second-order interaction to the data in Table 7.3*

Variable		Magnitude of effect	Standard error	Standardized value
1	$u_{1(L)}$	0.164	0.104	1.582
	$u_{1(Q)}$	0.012	0.059	0.201
2	$u_{2(1)}$	− 0.022	0.084	− 0.262
3	$u_{3(1)}$	− 0.215	0.123	− 1.754
	$u_{3(2)}$	0.665	0.101	6.569
	$u_{3(3)}$	− 0.450	0.130	− 3.452
1 and 2	$u_{12(L1)}$	0.145	0.104	1.394
	$u_{12(Q1)}$	0.061	0.059	1.044
2 and 3	$u_{23(11)}$	0.071	0.123	0.578
	$u_{23(12)}$	0.075	0.101	0.739
	$u_{23(13)}$	− 0.146	0.130	− 1.117
1 and 3	$u_{13(L1)}$	0.350	0.153	2.286
	$u_{13(L2)}$	0.257	0.125	2.057
	$u_{13(L3)}$	− 0.607	0.160	− 3.791
	$u_{13(Q1)}$	− 0.031	0.085	− 0.370
	$u_{13(Q2)}$	0.017	0.071	0.241
	$u_{13(Q3)}$	0.014	0.092	0.155

Table 7.5 *Number of people in the three categories of the duration variable, for the three levels of social class*

| | | Duration of present occupation (year) | | |
		< 2	2–5	5 +
	I + II	9	16	26
Social class	III	25	35	60
	IV + V	20	12	8

The more interesting effects are those representing interactions between duration of present occupation and diagnosis or social class. Examining Table 7.4 it is seen that interactions between duration and diagnosis are not significant. This indicates that the trends across duration are the same for both diagnoses. Interaction effects between duration and social class are, however, significant; the magnitude and sign of these effects show that there are **linear increases** in the number of people with increase in duration for social class categories 1 and 2, and a **linear decrease** for social class category 3. This is clearly seen in Table 7.5. The absence of any second-order relationship shows that the interaction effects between social class and duration of present occupation are the same for both neurotics and schizophrenics.

7.4 The linear-by-linear association model

In Chapter 5 the log-linear models considered for a two-dimensional contingency table were the independence model with $(r - 1)(c - 1)$ d.f. and the saturated model with zero d.f. A number of authors including Goodman (1979) and Agresti (1990) have considered a model which lies between these two and which has proved extremely useful for modelling the association between two ordinal variables. The model requires scores u_i and v_j to be assigned to the row and column categories (cf. section 7.1), where $u_1 \leqslant u_2 \leqslant \cdots \leqslant u_r$ and $v_1 \leqslant v_2 \leqslant \cdots \leqslant v_c$. The model proposed is then:

$$\ln F_{ij} = u + u_{1(i)} + u_{2(j)} + \beta u_i v_j. \tag{7.5}$$

The independence model is the special case, $\beta = 0$. Since the scores u_i and v_j are fixed, the model in (7.5) has only one more parameter, β, than the independence model; consequently it has degrees of freedom

$$\text{d.f.} = (r - 1)(c - 1) - 1 = rc - r - c. \tag{7.6}$$

The model requires only the single parameter β to describe association regardless of the number of rows and columns of the table, whereas the saturated model requires $(r - 1)(c - 1)$ parameters. The deviation, $\beta u_i v_j$, of $\ln F_{ij}$ from independence is linear in each variable when the other is regarded as fixed. Because of this property (7.5) is known as the **linear-by-linear association** model.

Interpretation of the magnitude of the parameter β in (7.5) is best made in terms of odds ratios. For any pair of columns $k < l$ and rows $s < t$ the linear-by-linear association model leads to

$$\ln\left(\frac{F_{sk}F_{tl}}{F_{sl}F_{tk}}\right) = \beta(u_t - u_s)(v_l - v_k). \tag{7.7}$$

The absolute log(odds ratio) is larger for pairs of rows or columns that are further apart. If **local odds ratios** are defined as

$$\theta_{ij} = \frac{F_{ij}F_{i+1,j+1}}{F_{i,j+1}F_{i+1,j}}, \tag{7.8}$$

then for the linear-by-linear association model

$$\ln \theta_{ij} = \beta(u_{i+1} - u_i)(v_{j+1} - v_j),$$

and for the equal-interval scores, all local odds ratios are equal. When the scores are unit-spaced, each local odds ratio is equal to $\exp(\beta)$. Goodman (1979) refers to this case as **uniform association**.

To illustrate the linear-by-linear association model, the data shown in Table 7.6 will be used. These data (taken with permission

Table 7.6   *Income and job satisfaction*

| Income($) | Job satisfaction | | | |
	Very dissatisfied	*Little dissatisfied*	*Moderately satisfied*	*Very satisfied*
< 6000	20	24	80	82
6000–15 000	22	38	104	125
15 000–25 000	13	28	81	113
> 25 000	7	18	54	92

Reproduced from Agresti (1990) permission of John Wiley and Sons.

from Agresti, 1990) concern a cross-classification of job satisfaction and income. Here the independence model gives a χ_L^2 value of 12.03 with nine d.f. ($p = 0.21$). Although this might often be regarded as adequate, in this case the examination of the adjusted residuals (see Table 7.7) indicates systematic lack of fit. The uniform association model has $\chi_L^2 = 2.39$ based on eight d.f. ($p = 0.97$), indicating a much improved fit. Examination of the fitted values from both models (see Table 7.8) shows that the uniform association model provides a much better fit in the corners of the table where it predicts the greatest departures from independence.

For unit-spaced scores the estimate of the association parameter is 0.112 with standard error 0.036. The positive value indicates that job satisfaction tends to be greater at higher income levels. The estimated uniform local odds ratio is $\hat{\theta}_{ij} = \exp(0.112) = 1.12$. A 95% confidence interval for θ_{ij} is $\exp(0.112 \pm 1.96 \times 0.036)$, i.e. (1.04, 1.20).

Table 7.7 *Adjusted residuals for job satisfaction data in Table 7.6*

| Income($) | Job satisfaction | | | |
	Very dissatisfied	*Little dissatisfied*	*Moderately satisfied*	*Very satisfied*
< 6000	4.44	− 0.17	1.17	− 1.94
6000–15 000	0.60	0.74	0.25	− 1.02
15 000–25 000	− 0.95	− 0.04	− 0.35	0.84
> 25 000	− 1.60	− 0.65	− 1.16	2.35

Table 7.8 *Fitted values for the independence and uniform association model for job satisfaction data*

| Income($) | Job satisfaction | | | |
	Very dissatisfied	*Little dissatisfied*	*Moderately satisfied*	*Very satisfied*
< 6000	14.2	24.7	72.9	94.2
	19.3	29.7	74.9	82.3
6000–15 000	19.9	34.6	102.3	132.2
	21.4	36.4	103.7	127.4
15 000–25 000	16.2	28.2	83.2	107.5
	13.6	25.9	82.4	113.2
> 25 000	11.8	20.5	60.5	78.2
	7.6	16.3	58.0	89.1

Top line for each salary group gives independence fitted values.

Similar models can be applied to tables of three and higher dimensions. For details see Agresti (1990, Ch. 8).

7.5 Linear-logistic models for ordered response variables

In the previous chapter linear-logistic models for categorical response variables with more than two categories were considered. The model outlined in section 6.4 is easily adapted to situations where the categories of the response variable may be considered ordered in some way. The model now assumes

$$Pr(y > j) = \frac{\exp v_j}{1 + \exp v_j} \tag{7.9}$$

for $j = 1, 2, \ldots, (J-1)$, $Pr(y > J) = 0$, and in this case $v_j = \alpha_j + \beta x$,

Table 7.9   *Observed cross-classification on 2294 young males who failed to pass the Armed Forces Qualification Test*

Race	Age	Father's education*	Respondent's education		
			Grammar school	Some HS	HS graduate
White	< 22	1	39	29	8
		2	4	8	1
		3	11	9	6
		4	48	17	8
	≥ 22	1	231	115	51
		2	17	21	13
		3	18	28	45
		4	197	111	35
Black	< 22	1	19	40	19
		2	5	17	7
		3	2	14	3
		4	49	79	24
	≥ 22	1	110	133	103
		2	18	38	25
		3	11	25	18
		4	178	206	81

*1 = Grammar school, 2 = Some HS, 3 = HS graduate, 4 = not available.
Reproduced from Fienberg (1980) by permission of the MIT Press.

where **x** is a vector of explanatory variables. Maximum likelihood methods are used to estimate the parameters.

To illustrate the use of this model the data shown in Table 7.9 (taken with permission from Fienberg, 1980), will be used. These data relate to a sample of 2400 young males rejected for military service because of failure to pass the Armed Forces Qualification Test. Here the respondent's education is considered to be an ordered response variable and race, age and father's education, categorical explanatory variables.

A stepwise procedure (Chapter 5) selects a model in which both race and father's education are included. The parameter estimates for such a model are shown in Table 7.10. (Father's education has been recoded in terms of three dummy variables – see Chapter 6.) Written out in full the model is as follows:

(1) Father's education–grammar school:

$$Pr(\text{respondent's education} > \text{grammar}) = \frac{\exp(-0.088 + D_1)}{1 + \exp(-0.088 + D_1)},$$
(7.10)

$$Pr(\text{respondent's education} > \text{some HS}) = \frac{\exp(-1.954 + D_1)}{1 + \exp(-1.954 + D_1)}.$$
(7.11)

(2) Father's education–some HS:

$$Pr(\text{respondent's education} > \text{grammar}) = \frac{\exp(-0.088 + D_2)}{1 + \exp(-0.088 + D_2)},$$
(7.12)

$$Pr(\text{respondent's education} > \text{some HS}) = \frac{\exp(-1.954 + D_2)}{1 + \exp(-1.954 + D_2)}.$$
(7.13)

(3) Father's education–HS graduate:

$$Pr(\text{respondent's education} > \text{grammar}) = \frac{\exp(-0.088 + D_3)}{1 + \exp(-0.088 + D_3)},$$
(7.14)

$$Pr(\text{respondent's education} > \text{some HS}) = \frac{\exp(-1.954 + D_3)}{1 + \exp(-1.954 + D_3)}.$$
(7.15)

Table 7.10 Parameter estimates for logistic regression model applied to data in Table 7.9

All outcomes	Coefficient	Standard error	Coefficient/SE	Exp(coefficient)	95% Confidence interval	
					Low	High
Race	0.84	0.08	10.50	2.32	2.18	2.71
Father's education (1)	0.55	0.17	3.32	1.73	1.23	2.42
(2)	1.07	0.15	7.02	2.91	2.17	3.91
(3)	−0.28	0.09	−3.26	0.75	0.64	0.89
X_1 (grammar)	−0.09	0.07	−1.16	0.92	0.79	1.05
X_2 (some HS)	−1.95	0.09	−21.67	0.14	0.12	0.17

(4) Father's education–not available:

$$Pr(\text{respondent's education} > \text{grammar}) = \frac{\exp(-0.088 + D_4)}{1 + \exp(-0.088 + D_4)},$$
$$(7.16)$$

$$Pr(\text{respondent's education} > \text{some HS}) + \frac{\exp(-1.954 + D_4)}{1 + \exp(-1.954 + D_4)}.$$
$$(7.17)$$

Where

$$D_1 = 0.839 \times \text{race}$$
$$D_2 = 0.839 \times \text{race} + 0.550$$
$$D_3 = 0.839 \times \text{race} + 1.073$$
$$D_4 = 0.839 \times \text{race} - 0.283$$

with race $= 0$ for whites and race $= 1$ for blacks. Such a model does not fit the data very well ($X^2 = 71.19$, $p < 0.001$), but as an exercise it is useful to try to interpret the coefficients in the model and what they imply about the data.

Taking first the coefficient for race; since the coefficient is positive it implies that, for a given level of father's education, a black respondent's probability of having some high school education rather than grammar, or being a high school graduate rather than having only some high school education is greater than for white respondents. Moreover, this effect can be quantified by exponentiating the coefficient to give the value 2.3 with a confidence interval of 2.0 to 2.7.

The coefficients for father's education demonstrate that for a given type of respondent (white or black), an increase in father's educational level increases the probability that the respondent will have a higher educational level. The coefficient for father's education not available is negative, indicating that here the probability of higher educational levels for the respondent are decreased. Again, the effects can be quantified by exponentiating the respective coefficients. For example, when the father is a high school graduate, the probability of a respondent being a high school graduate rather than simply having had some high school education is 2.9 times that of a respondent whose father was grammar school educated.

7.6   Association measures for tables with ordered categories

In this section we shall discuss measures of association that are specifically designed for the situation where the variables forming the contingency table have ordered categories. Such measures will take positive values when 'high' values of one variable tend to occur with 'high' values of the other variable, and 'low' with 'low'. In the reverse situation, the coefficients will be negative. One obvious method for obtaining a measure of association for ordered tables would be to assign scores to the categories and then compute the product–moment correlation coefficient between the two variables. A difficulty arises, however, in deciding on the appropriate scoring system to use. Many investigators would be unhappy about imposing a metric on the categories in their table, and consequently require a measure of association which does not depend on imposing a set of arbitrary scores. Here three such measures will be discussed, namely the tau statistics of Kendall, Goodman and Kruskal's gamma, and Somer's d.

7.6.1   Kendall's tau statistics

Kendall's tau, τ, is well known as a measure of correlation between two sets of rankings. It may be adapted for the general $r \times c$ contingency table having ordered categories by regarding the table as a way of displaying the rankings of the N individuals according to two variables, for one of which only r separate ranks are distinguished and for the other only c separate ranks are distinguished. From this point of view the marginal frequencies in the table are the number of observations 'tied' at the different rank values distinguished. Kendall's tau measures are based on S which is given by:

$$S = P - Q, \tag{7.18}$$

where P is the number of concordant pairs of observations, that is pairs of observations such that their rankings on the two variables are in the same direction, and Q is the number of discordant pairs for which rankings on the two variables are in the reverse direction. (For computation of S see later.) To obtain a measure of association from S it must be standardized to lie in the range from -1 to $+1$. Different methods of standardization give rise to three different tau statistics:

$$\tau_a = \frac{2S}{N(N-1)}, \tag{7.19}$$

$$\tau_b = \frac{2S}{\sqrt{[(P+Q+X_0)(P+Q+Y_0)]}}, \tag{7.20}$$

where X_0 represents the number of observations tied to the first variable only, and Y_0 the number of observations tied on the second variable only (again for computation see later), and

$$\tau_c = \frac{2mS}{N^2(m-1)}, \tag{7.21}$$

where $m = \min(r,c)$.

τ_a is the commonly used measure of rank correlation. It is not applicable to contingency table data since it assumes that there are no tied observations. The other two coefficients may, however, be used to measure association in ordered tables. Kendall *et al.* (1987, Vol. 2, Ch. 33) show that τ_b may only attain the values $+1$ for a square table, but that τ_c can reach these extreme values apart from a small effect produced when N is not a multiple of m. The main problem with these two measures is that they have no obvious probabilistic interpretation, and consequently the meaning of a value of τ_b of 0.7 or τ_c of 0.6, say, cannot be expressed in words in terms of probabilities or errors in prediction.

7.6.2 *Goodman and Kruskal's gamma*

Goodman and Kruskal suggest a measure of association for ordered tables also based on S and given by:

$$\gamma = \frac{S}{P+Q}. \tag{7.22}$$

This coefficient has the considerable advantage of having a direct probabilistic interpretation, namely as the difference in probability of like rather than unlike orders for the two variables when two individuals are chosen at random. γ takes the value $+1$ when the data are concentrated in the upper-left to lower-right diagonal (assuming that both variables are ordered in the same direction either both 'low' to 'high' or both 'high' to 'low'). It takes the value zero in the case of independence, but the converse need not hold.

7.6.3 Somers's d

Somers (1962) gives a measure of association for contingency tables with ordered categories which is suitable for the asymmetric case in which we have an explanatory and a dependent variable. This coefficient d_{yx} is given by:

$$d_{yx} = \frac{S}{P + Q + Y_0}, \qquad (7.23)$$

where x indicates the explanatory and y the dependent variable. In this case Y_0 represents the number of observations tied on the dependent variable. This coefficient has a similar interpretation to that of γ. By noting the relationship

$$\tau_b^2 = d_{yx}d_{xy} \qquad (7.24)$$

it is seen that the ds bear the same relationship to Kendall's correlation measure as the classical regression coefficients bear to the product moment correlation coefficient, namely $r^2 = b_{yx}b_{xy}$. Somers's d coefficients can therefore be thought of as analogous to the ordinary regression coefficients.

7.6.4 Numerical example illustrating the computation of the τ statistics, γ and d

The data in Table 7.11 were collected during an investigation into attempted suicides, and show suicidal intent and a depression rating score for a sample of 91 cases. The data are taken from Birtchnell and Alarcon (1971). We begin by computing P and Q as follows.

P: each cell in the table is taken in turn and the number of observations in the cell is multiplied by the number of observations in each cell to its south-east and the terms are summed. Cells in the

Table 7.11   *Suicidal intent and depression rating score*

		Depression rating (x)				
		<20	20–29	30–39	>39	
Suicidal	Did not want to die	10	14	8	2	34
intent (y)	Unsure	2	4	7	2	15
	Wanted to die	5	9	11	17	42
		17	27	26	21	91

same row and column are ignored. For the data of Table 7.10 then we have:

$$P = 10(4 + 7 + 2 + 9 + 11 + 17) + 14(7 + 2 + 11 + 17)$$
$$+ 8(2 + 17) + 2(9 + 11 + 17) + 4(11 + 17) + 7(17)$$
$$= 1475.$$

Q: each cell is taken in turn and the number of observations in the cell is multiplied by the number of observations in each cell to its south-west. Again, cells in the same row and column are ignored. Then for Table 7.11 we have:

$$Q = 14(2 + 5) + 8(2 + 4 + 5 + 9) + 2(2 + 4 + 7 + 5 + 9 + 11)$$
$$+ 4(5) + 7(5 + 9) + 2(5 + 9 + 11)$$
$$= 502.$$

Now both X_0 and Y_0, the number of observations tied on the depression rating only (the 'x' variable) and the number tied only on suicidal intent (the 'y' variable), need to be calculated. This may be done as follows.

$X_0(Y_0)$: each cell is taken in turn and the number of observations in the cell is multiplied by the number of observations following it in the particular column (row) involved, and the results are summed. For Table 7.11 we have:

$$X_0 = 10(2 + 5) + 2(5) + 14(4 + 9) + 4(9) + 8(7 + 11)$$
$$+ 7(11) + 2(2 + 17) + 2(17)$$
$$= 591.$$

Similarly,

$$Y_0 = 10(14 + 8 + 2) + 14(8 + 2) + 8(2) + 2(4 + 7 + 2)$$
$$+ 4(7 + 2) + 7(2) + 5(9 + 11 + 17) + 9(11 + 17) + 11(17)$$
$$= 1096.$$

Having obtained P, Q, X_0 and Y_0, the τ statistics, γ and d may be calculated as follows:

$$\tau_b = \frac{2(1475 - 502)}{\sqrt{[(1475 + 502 + 591)(1475 + 502 + 1096)]}}$$
$$= 0.69,$$

$$\tau_c = \frac{2 \times 3(1475 - 502)}{91^2 \times 2}$$
$$= 0.35,$$

$$\gamma = \frac{1475 - 502}{1475 + 502}$$

$$= 0.49,$$

$$d_{yx} = \frac{1475 - 502}{1475 + 502 + 1096}$$

$$= 0.32.$$

It is seen that the values of the coefficients differ considerably, but all indicate considerable positive association between depression rating score and suicidal intent. In other words, patients having high depression scores tended to be those who expressed the view that they 'wished to die' in their suicide attempt.

Because of its clear probabilistic interpretation the coefficient γ is perhaps the most useful for measuring association in ordered tables.

7.6.5   Polychoric and polyserial correlation

The **polyserial** and **polychoric** correlations are measures of bivariate association arising when one or both observed variates are ordered, categorical variables that result from grouping one or two underlying continuous variables. The categorical variables, A and B, are

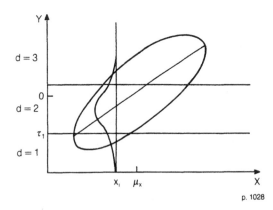

p. 1028

Fig. 7.1. *Example of two ordered, categorical variables formed by imposing thresholds on underlying continuous variables. Reproduced from the* Encyclopedia of Statistical Science, *Vol. 5, by permission of John Wiley and Sons.*

assumed to be related to underlying continuous variables, X and Y, by

$$A = a_i \text{ if } \gamma_{i-1} \leqslant X < \gamma_i \, i = 1,2,\ldots,r,$$

and

$$B = b_j \text{ if } \tau_{j-1} \leqslant Y < \tau_j \, j = 1,2,\ldots,c.$$

The parameters γ_i and τ_j are known as **thresholds**. The process is illustrated in Figure 7.1. When $r = c = 2$ the relevant coefficient is known as the **tetrachoric correlation** which was first studied by Pearson (1909) who gave a series of approximations for the measure including

$$Q_1 = \frac{ad - bc}{ad + bc},$$

$$Q_2 = \sin\frac{\pi}{2} \frac{\sqrt{ad} - \sqrt{bc}}{\sqrt{ad} + \sqrt{bc}},$$

$$Q_3 = \sin\frac{\pi}{2}\left[1 + \frac{2bcN}{(ad - bc)(b + c)}\right]^{-1}.$$

Such approximations are, however, rarely needed nowadays since all these coefficients can be relatively easily found by maximum likelihood methods. Details are to be found in Olsson *et al.* (1982) and Lee and Poon (1986).

7.7 Summary

Many models for contingency tables with invariables having ordered categories have been developed during the last decade. Only a small number have been described in this chapter and the description has been of necessity relatively brief. More detailed accounts of the area are available in Agresti and Kezouh (1983), Becker and Clogg (1989) and Agresti (1983, 1984).

CHAPTER 8

Some special types of contingency table

8.1 Introduction

There are certain types of contingency table met with in practice that require special consideration. In this chapter a number of methods of analysis which have been suggested for such tables are discussed, beginning with a description of the analysis of tables containing *a priori* zero entries.

8.2 Tables with *a priori* zeros

In Chapter 5 the problem of contingency tables having a number of empty cells was mentioned. In the case of **sampling zeros** the solution was either to increase the sample size or, failing this, to add a small positive constant (for example, 0.5) to each cell frequency. In many situations, however, tables arise in which it is theoretically impossible to have observations in a cell; in this case the empty cells are usually referred to as **structural zeros**, and the table as a whole as **incomplete**. The analysis of such tables presents special problems which have been considered by several authors, including Goodman (1968), Mantel (1970), Fienberg (1972) and Bishop *et al.* (1975, Ch. 5). Since much of this work is outside the scope of this text, it will simply be illustrated by means of an example; for this purpose the data in Table 8.1 will be used. (These data were originally given by Brunswick, 1971, and also considered by Grizzle and Williams, 1972.) The data involve health problems causing concern amongst a sample of teenagers. Since males are naturally not affected by menstrual problems certain cells contain *a priori* zeros.

The analysis of such a table involves the use of log-linear models from which parameters referring to cells containing the structural zeros are excluded since they are known *a priori* to be zero. Expected values for such models may be obtained by using a simple modification of the algorithm described in section 5.5; starting values for this algorithm are now taken as unity for non-empty cells, and **zero** for the *a priori* empty cells. Using such starting values ensures that the table of expected values corresponding to a particular model will have zeros in the cells required. The observed and expected frequencies are then compared by means of the familiar X_L^2 statistic. The calculation of the correct degrees of freedom for the test is, however, complicated by the presence of the structural zeros. Previously degrees of freedom have been found from the equation (see equation (5.34)):

$$\text{d.f.} = N_1 - N_2, \tag{8.1}$$

where N_1 = number of cells in the table and N_2 = number of parameters in the model that need to be estimated. In the case of incomplete tables the corresponding formula is:

$$\text{d.f.} = N_1 - N_2 - N_3, \tag{8.2}$$

where N_3 = number of *a priori* empty cells. Care is needed in determining the number of parameters to be estimated, since those referring to the empty cells are known *a priori* to be zero and must therefore be excluded.

Returning to the data in Table 8.1, consider first the mutual independence model, for which the expected values are shown in Table 8.2. The statistic X_L^2 takes the value 28.24. The degrees of freedom in this case are

$$16 - 2 - (1 + 1 + 1 + 3) = 8$$

since there are two empty cells and the parameters that have to be estimated are u (grand mean), $u_{1(1)}$ (age effect), $u_{2(1)}$ (sex effect) and

Table 8.1 *Data on teenagers' concern with health problems*

		Males		Females	
		Sex (variable 2)			
Age (variable 1)		12–15	16–17	12–15	16–17
	S	4	2	9	7
Health problem	M	–	–	4	8
causing concern	H	42	7	19	10
(variable 3)	N	57	20	71	31

S: Sex reproduction.   H: How healthy I am.
M: Menstrual problems.   N: Nothing.

Table 8.2 *Expected values under the complete independence model for the data shown in Table 8.1*

| | | Sex | | | |
		Males		Females	
Age		12–15	16–17	12–15	16–17
Health problem	S	7.37	3.04	8.21	3.39
causing concern	M	–	–	8.49	3.51
	H	26.12	10.78	29.09	12.00
	N	59.95	24.74	66.76	27.55

$u_{3(1)}$, $u_{3(2)}$, $u_{3(3)}$ (health concern effects). Clearly the model does not fit adequately.

For a model specifying only that there is no second-order interaction, a value of $X_L^2 = 2.03$ is obtained. To determine the degrees of freedom, first consider the parameters in the model that have to be estimated, remembering that the parameters are such that $u_{1(.)} = 0$, $u_{2(.)} = 0$, $u_{12(.1)} = 0$ etc.

Grand mean effect	u
Main effect age	$u_{1(1)}$
Main effect, sex	$u_{2(1)}$
Main effect, health concern	$u_{3(1)}, u_{3(2)}, u_{3(3)}$
Interaction effect, age × sex	$u_{12(11)}$
Interaction effect, age × concern	$u_{13(11)}, u_{13(12)}, u_{13(13)}$
Interaction effect, sex × concern	$u_{23(11)}, -, u_{23(13)}$

The interaction effect, $u_{23(12)}$, is not needed because of the corresponding structural zeros. The number of parameters to be estimated is therefore 12, and the degrees of freedom is given by

$$16 - 2 - 12 = 2.$$

Clearly no second-order interaction needs to be postulated for these data.

8.3 Quasi-independence

The fitting of models to tables with *a priori* zeros is closely related to the problem of examining contingency tables for what is termed by Goodman (1968) **quasi-independence**, and which Fienberg (1972) has described as a form of independence **conditional** on restricting attention to a particular part of the table only. Testing tables for independence is often a useful way of identifying the sources of significance in tables where the overall test statistic is significant. To illustrate the concept, the data in Table 8.3 will be used. These data were collected by Glass (1954) in a study of social mobility in Great Britain.

The usual chi-square test of independence, that is of the hypothesis

$$H_0 : p_{ij} = p_{i.}p_{.j}$$

gives $X^2 = 505.5$, which with four d.f. indicates that the two classifications are not independent. (The expected frequencies under the hypothesis that father's and son's classes are independent are shown in parentheses in Table 8.3.)

For the data in Table 8.3 hypotheses other than that of simple independence might be of more interest. For example, suppose interest centred on investigating whether the data are compatible with the theory that, whilst there may be some 'class inheritance' from father to son in every social stratum, once a son has moved out of his father's particular class, his destination class is independent of that of his father. Testing such a theory would entail testing for independence in that portion of the table given by excluding sons having the same status as their fathers, i.e. excluding the main diagonal of the table; Table 8.4. This latter table may be thought of as arising from sampling a form of truncated population from which sons having the same class as their fathers have been excluded. The usual form of the hypothesis of independence given above must now be modified to reflect the fact that the diagonal elements of Table 8.4 are blank, by setting $p_{ii} = 0$ and adjusting the remaining probabilities

Table 8.3 *Cross-classification of a sample of British males according to each subject's status category and his father's status category*

		Upper	Subject's status Middle	Lower
		Upper	Middle	Lower
	Upper	588 (343.2)	395 (466.7)	159 (322.1)
Father's status	Middle	349 (453.8)	714 (617.0)	447 (439.1)
	Lower	114 (254.0)	320 (345.3)	411 (245.7)

The expected values under the hypothesis that a subject's status is independent of his father's status are shown in parentheses alongside the observed values.

Table 8.4   *Data of Table 8.3 with diagonal elements excluded*

		\-	Subject's status	\-
		Upper	Middle	Lower
Father's status	Upper	–	395	159
	Middle	349	–	447
	Lower	114	320	–

so that they continue to sum to 1; the hypothesis now takes the form

$$H_0 : p_{ij} = 0 \text{ if } i = j$$
$$= S p_{i.} p_{.j} \text{ if } i \neq j,$$

where S is chosen so that:

$$\sum_{i=1}^{3} \sum_{j=1}^{3} p_{ij} = 1,$$

that is

$$S = \left(1 - \sum_{i=1}^{3} p_{i.} p_{.j} \right)^{-1}.$$

This is now termed a hypothesis of quasi-independence. The maximum likelihood estimators of $p_{i.}$ and $p_{.j}$, and consequently of the expected frequencies, cannot be found directly. The latter can, however, be found by using the iterative algorithm described in Chapter 5, in a manner similar to that outlined in the preceding section, and using starting values of unity for the cells of interest and zero for those to be excluded. In the social mobility example this leads to

$$E_{ij}^{(0)} = 1 \text{ if } i \neq j$$
$$= 0 \text{ if } i = j,$$

where the $E_{ij}^{(0)}$ are the starting values for the algorithm. Once the expected values under the hypothesis of quasi-independence have been found they are compared to the observed frequencies by means of the usual X^2 of X_L^2 statistics.

The expected values for the data in Table 8.4 are shown in Table 8.5; these lead to a value of $X^2 = 0.61$ with a single degree of freedom (due to subtraction of three d.f. for the diagonal cells). Clearly

Table 8.5 *Estimated expected values under the hypothesis of quasi-independence, excluding subjects who have the same status as their fathers*

		Subject's status		
		Upper	Middle	Lower
	Upper	–	390.2	163.8
Father's status	Middle	353.8	–	442.2
	Lower	109.2	324.8	–

the test statistic is not significant and so the data are compatible with the previously described theory of social mobility.

The small value of the chi-square test statistic for quasi-independence demonstrates that the large value found when testing independence in the complete table is due to discrepancies between the observed and expected values in those cells where father and son have the same class. In each of these the observed frequency is greater than would be expected if the two classifications were independent. The greatest discrepancy is for the cell involving upper class sons and upper class fathers, and it might be of interest to test for quasi-independence when only this diagonal cell is excluded. In this case the following initial values would be suitable for the iterative algorithm.

$$E_{ij}^{(0)} = 0 \text{ if } i = j = 1,$$

$$= 1 \text{ otherwise.}$$

The resulting expected values lead to a value $X^2 = 143.42$, which with three d.f. is highly significant; consequently 'status inheritance' does not occur only in the upper class structure.

Proceeding now one stage further and excluding also lower class sons having a lower class father, leads to $X^2 = 20.20$ which with two d.f. is also significant. From these results it would appear that 'status inheritance' occurs in all three social strata.

Agresti (1990) shows that the quasi-independence model has equivalent log-linear form:

$$\log F_{ij} = u + u_{1(i)} + u_{2(j)} + \delta_i I(i = j), \tag{8.3}$$

where I is a function defined as follows

$$I(i = j) = 1 \text{ if } i = j$$

$$= 0 \text{ if } i \neq j.$$

The first three terms in the model specify independence, and the $[\delta_i]$ parameters permit the F_{ii}, i.e. the frequencies in the diagonal cells, to depart from this pattern.

8.4 Square contingency tables

Two-dimensional contingency tables in which the row and column variables have the same number of categories (say r) occur fairly frequently in practice and are known in general as **square tables**. They may arise in a variety of ways:

(1) When a sample of individuals is cross-classified according to two essentially similar categorical variables, for example grade of vision of left and right eye.
(2) When samples of pairs of matched individuals, such as husbands and wives, fathers and sons, are each classified according to some categorical variable of interest (in the case of a 2×2 table this situation was considered in section 2.2).
(3) When two raters independently assign a sample of subjects to a set of categories.

For such tables hypotheses relating simply to independence are not of major importance. Instead interest centres on testing for **symmetry** and **marginal homogeneity**. By symmetry in a square table is meant

$$p_{ij} = p_{ji} \quad (i \neq j), \tag{8.4}$$

and by marginal homogeneity that

$$p_{i.} = p_{.j} \quad (\text{for } i = 1, 2 \ldots r). \tag{8.5}$$

In a 2×2 table these are obviously equivalent; in larger tables symmetry as defined by (8.4) clearly implies marginal homogeneity as defined by (8.5). Chi-square tests for both symmetry and marginal homogeneity are available and are described in the next two sections.

8.4.1 Testing for symmetry

Examining again the data in Table 8.3, a further question of interest might be whether or not changes in class between fathers and sons occur in both directions with the same probability, i.e. to determine whether observations in cells situated symmetrically about the main diagonal have the same probability of occurrence. This hypothesis of symmetry

$$H_0 : p_{ij} = p_{ji} \,(i \neq j)$$

has been considered by several authors including Grizzle *et al.* (1969), Maxwell (1970), Bishop *et al.* (1975, Ch. 8) and Agresti (1990, Ch. 10). Under such a hypothesis, the frequencies in the *ij*th and *ji*th cells are expected to be equal and the maximum likelihood estimate is:

$$E_{ij} = \tfrac{1}{2}(n_{ij} + n_{ji})\,(i \neq j),$$

$$= n_{ii}\,(i = j). \tag{8.6}$$

Substituting these values in the usual form of the chi-square test statistic gives:

$$X^2 = \sum_{i<j} (n_{ij} - n_{ji})^2 / (n_{ij} + n_{ji}), \tag{8.7}$$

which under the hypothesis of symmetry has a chi-square distribution with $\tfrac{1}{2}r(r-1)$ d.f. For the data in Table 8.3:

$$X^2 = \frac{(349 - 395)^2}{(349 + 395)} + \frac{(159 - 114)^2}{(159 + 114)} + \frac{(447 - 320)^2}{(447 + 320)}$$

$$= 2.84 + 7.42 + 21.03$$

$$= 31.29.$$

This has three degrees of freedom and is highly significant; consequently the hypothesis of symmetry is rejected. The largest deviation occurs between frequencies n_{32} and n_{23}, and it appears that a larger number of sons of middle class fathers become lower class than the number of sons of lower class fathers who achieve middle class status; to a lesser extent the same is true for the number of sons of upper class fathers who become lower class, which is greater than the number of sons who go in the opposite direction.

Agresti (1990) shows that symmetry has the following log-linear representation:

$$\log F_{ij} = u + u_{1(i)} + u_{2(j)} + u_{12(ij)}, \tag{8.8}$$

where $u_{12(ij)} = u_{12(ji)}$ and both classifications have the same single-factor parameters.

8.4.2 Testing for marginal homogeneity

In cases where the hypothesis of symmetry is rejected, the weaker hypothesis of marginal homogeneity may be of interest. This hypothesis corresponds to equality of row and column marginal probabilities and may be written as:

$$H_0 : p_{i.} = p_{.i} (i = 1, 2, ..., r)$$

A suitable test for this hypothesis has been given by Stuart (1955) and by Maxwell (1970). For the general $r \times r$ table the test statistic is given by:

$$X^2 = \mathbf{d}'\mathbf{V}^{-1}\mathbf{d}, \tag{8.9}$$

where $\mathbf{d}$ is a column vector of any $(r-1)$ of the differences $d_1, d_2, ..., d_r$ where $d_i = n_{i.} - n_{.i}$ and $\mathbf{V}$ is the $(r-1) \times (r-1)$ matrix of the corresponding variance and covariances of the d_i, having elements:

$$v_{ii} = n_{i.} + n_{.j} - 2n_{ij}, \tag{8.10}$$

$$v_{ij} = -(n_{ij} + n_{ji}). \tag{8.11}$$

(Stuart and Maxwell show that the same value of X^2 results whichever value of d_i is omitted from the vector $\mathbf{d}$.)

If H_0 is true then the test statistic given in (8.9) has a chi-square distribution with $(r-1)$ d.f.

Fleiss and Everitt (1971) show that when $r = 3$ the test statistic may be written as

$$\chi^2 = \frac{\bar{n}_{23}d_1^2 + \bar{n}_{13}d_2^2 + \bar{n}_{12}d_3^2}{2(\bar{n}_{12}\bar{n}_{23} + \bar{n}_{12}\bar{n}_{13} + \bar{n}_{13}\bar{n}_{23})}, \tag{8.12}$$

where

$$\bar{n}_{ij} = \frac{1}{2}(n_{ij} + n_{ji}).$$

Applying (8.12) to the data in Table 8.3 for which $d_1 = -91$, $d_2 = -81$, $d_3 = 17$, and $\bar{n}_{12} = 372.0$, $\bar{n}_{13} = 136.5$; $\bar{n}_{23} = 383.5$, gives $X^2 = 30.67$, which with two d.f. is highly significant. Clearly the marginal distribution of father's status differs from that of their sons. The relevant marginal probabilities estimated from Table 8.4 are as follows:

		Class status	
	Upper	Middle	Lower
Father's	0.33	0.43	0.24
Son's	0.30	0.41	0.29

The results indicate a general lowering in the class status of sons compared with that of their fathers.

Other tests of marginal homogeneity are given in Bishop *et al.* (1975, Ch. 8); these authors also consider the concepts of symmetry and marginal homogeneity in the case of tables with more than two dimensions.

8.5 Square tables with ordered categories

Several specialized models have been developed to investigate conditions analogous to symmetry and marginal homogeneity for square tables with ordered categories. Many of these are described in Agresti (1984); an example of a model first given by McCullagh (1978) will serve to illustrate the possibilities.

The data shown in Table 8.6 describe unaided distance vision for a sample of women. A simple test of symmetry as described previously gives $X^2 = 19.106$ with six d.f. and an associated p value of 0.004; clearly the model does not fit these data at all well. McCullagh (1978) proposes an alternative to symmetry, known as the **conditional symmetry** model, applicable as in this example, when the variables forming the table are ordered. The hypothesis of interest is now:

$$H_0: P_{ij} = P_{ji} \quad i < j,$$

where

$$P_{ij} = Pr(X = i, Y = j / X < Y)$$
$$P_{ji} = Pr(X = j, Y = i / X > Y) \tag{8.13}$$

with X and Y representing the variables forming the table.

Such a model implies that the frequencies above the main diagonal are some constant times the frequencies below the main diagonal, i.e.

$$F_{ij} = \delta F_{ji} \quad i < j. \tag{8.14}$$

Agresti (1984) gives the following estimators for the constant and the frequencies, F_{ij}:

$$\hat{\delta} = \frac{\Sigma_{i<j}\Sigma n_{ij}}{\Sigma_{i>j}\Sigma n_{ij}}, \tag{8.15}$$

$$E_{ij} = \frac{\hat{\delta}(n_{ij} + n_{ji})}{(\hat{\delta} + 1)} \quad i < j, \tag{8.16}$$

$$E_{ij} = \frac{(n_{ij} + n_{ji})}{(\hat{\delta} + 1)} \quad i > j. \tag{8.17}$$

Table 8.6   *Unaided distance vision of 7477 women aged 30–9 employed in Royal Ordnance factories from 1943 to 1946*

Right eye		Highest grade (1)	Left eye (2)	(3)	Lowest grade (4)	Total
Highest	(1)	1520	266	124	66	1976
	(2)	234	1512	432	78	2256
	(3)	117	362	1772	205	2456
Lowest grade	(4)	36	82	179	492	789
Total		1907	2222	2507	841	7477

Reported from McCullagh (1978) by permission of Biometrika.

The conditional symmetry model has one additional parameter over the symmetry model, so has $r(r-1)/2 - 1 = (r+1)(r-2)$ with two d.f. For the data in Table 8.6 the estimate of δ is 1.16 and the value of X^2 is 7.26 which with five d.f. has an associated p value of 0.20. The conditional symmetry model fits the data adequately and the value of $\hat{\delta}$ suggests that the vision in the left eye is rather poorer than in the right.

8.6   Measures of agreement

A common situation in which square contingency tables arise is when two observers separately classify a sample of subjects using the same categorical scale. For example, two psychiatrists might classify psychiatric patients into a number of prespecified diagnostic categories. The joint ratings of the two clinicians can be displayed in a square table, with the same categories for rows and columns. Table 8.7 shows an example.

Such data are often used to investigate the reliability of the categorical scale, usually by evaluating agreement between the two observers. An intuitively appealing index of agreement is simply the proportion of patients classified into the same category by the two investigators, that is P_o given by

$$P_o = \sum_{i=1}^{r} n_{ii}/N, \qquad (8.18)$$

which for Table 8.7 gives $P_o = 0.63$. Such a measure has the virtue

Table 8.7 *Psychiatrists' diagnosis of 118 patients*

		Psychiatrist 1				
		D1	D2	D3	D4	D5
	D1	22	2	2	0	0
	D2	5	7	14	0	0
Psychiatrist 2	D3	0	2	36	0	0
	D4	0	1	14	7	0
	D5	0	0	3	0	3

of simplicity and is also readily understand. It is not, however, an adequate measure of agreement since it ignores agreement between the observers that might be due to chance. To illustrate the problem consider Tables 8.8 and 8.9.

In both these tables P_o given by (8.18) is 0.66. But in Table 8.8 this level of agreement could have been produced by the observers simply allocating subjects to categories **at random** in accordance with their particular marginal rates. For example, observer 1 would simply allocate 10% of the subjects to category 1, 80% to category 2 and the remaining 10% to category 3 **without** regard to which patients were involved. The level of agreement produced by such an allocation process can be calculated as:

$$P_c = \frac{1}{N}\left(\sum_{i=1}^{r} n_{i.}n_{.i}/N \right), \tag{8.19}$$

giving for Table 8.8:

$$P_c = \frac{1}{100}\left(\frac{10 \times 10}{100} + \frac{80 \times 80}{100} + \frac{10 \times 10}{100} \right) = 0.66.$$

So in this particular table all the observed agreement might simply be due to chance. On the other hand, the agreement to be expected

Table 8.8 *Hypothetical data showing 66% agreement between two observers*

		Observer 1		
		1	2	3
	1	1	8	1
Observer 2	2	8	64	8
	3	1	8	1

Table 8.9   *Further set of data showing 66% agreement between two observers*

		Observer 1		
		1	2	3
	1	24	13	3
Observer 2	2	5	20	5
	3	1	7	22

by chance in Table 8.9 is given by:

$$P_c = \frac{1}{100}\left(\frac{40 \times 30}{100} + \frac{30 \times 40}{100} + \frac{30 \times 30}{100}\right) = 0.33$$

which is considerably lower than the observed agreement.

Several authors have expressed opinions on the need to incorporate chance agreement into the assessment of inter-observer reliability. The clearest statement in favour of such a correction has been made by Fleiss (1975), who suggests the following procedure as a natural means of correcting for chance agreement.

When there is complete agreement between the two observers, P_o will take the value 1, so that the maximum possible excess over chance agreement is $1 - P_c$. The observed excess over chance agreement is $P_o - P_c$ and the ratio of the two differences, namely

$$\kappa = \frac{P_o - P_c}{1 - P_c}, \tag{8.20}$$

is a measure of agreement with several attractive properties. If there is a complete agreement, $\kappa = 1$. If observed agreement is greater than chance then $\kappa > 1$. If observed agreement is equal to chance, $\kappa = 0$. Finally if the observed agreement is less than chance, $\kappa < 0$, with its minimum value depending on the marginal distributions.

The chance-corrected measure of agreement given in (8.20) was first suggested by Cohen (1960) and is generally known as the **kappa** coefficient. For the data in Table 8.7 the coefficient is given by

$$\kappa = \frac{0.63 - 0.27}{1 - 0.27} = 0.49.$$

Landis and Koch (1977) give some arbitrary but potentially useful 'benchmarks' for evaluating observed values of the kappa coefficient.

They are as follows:

κ	Strength of Agreement
0.00	Poor
0.00–0.20	Slight
0.21–0.40	Fair
0.41–0.60	Moderate
0.61–0.80	Substantial
0.81–1.00	Almost perfect

For some researchers, informal evaluation of observed kappa values will not be sufficient; instead they will be interested in testing hypotheses about population kappas, for example that kappa = 0, or in testing whether there is a difference in two kappa values. Additionally, confidence intervals for kappa may be required. Consequently, the variance of a sample kappa may be required, and this has been derived under a number of assumptions by Everitt (1968) and Fleiss et al. (1969). The asymptotic large sample variance of κ may be estimated from:

$$\text{Var}(\kappa) = \frac{1}{n(1 - P_c)^4} \left\{ \sum_{1=1}^{r} p_{ii}[(1 - P_c) - (p_{.i} + p_{i.})(1 - P_o)] \right.$$
$$\left. + (1 - P_o)^2 \sum_{\substack{i=1 \\ i \neq j}}^{r} \sum_{j=1}^{r} p_{ij}(p_{.i} + p_{j.})^2 - (P_o P_c - 2P_c + P_o)^2] \right\}$$

(8.21)

where p_{ij} represents the proportion of observations in the ijth cell of the table and $p_{i.}$ and $p_{.j}$ are row and column marginal proportions.

To illustrate the use of the rather horrendous expression in (8.21), the data in Table 8.7 will be used. The various terms needed for the calculation of the variance are given in Table 8.10 and lead to:

$$\text{Var}(\kappa) = 0.0036.$$

So the estimated standard deviation of κ is 0.06. An approximate 95% confidence interval is (0.37, 0.61).

The concept of a chance-corrected measure of agreement can be extended to situations involving more than two observers; for details see Fleiss and Cuzick (1979) and Schouten (1985).

Table 8.10 *Psychiatric diagnosis: terms needed for the calculation of the variance of κ*

		Psychiatrist 1					
		D1	D2	D3	D4	D5	
Psychiatrist 2	D1	0.186	0.017	0.017	0.000	0.000	0.220
	D2	0.042	0.059	0.119	0.000	0.000	0.220
	D3	0.000	0.017	0.305	0.000	0.000	0.322
	D4	0.000	0.008	0.119	0.059	0.000	0.186
	D5	0.000	0.000	0.025	0.000	0.025	0.051
		0.229	0.102	0.585	0.059	0.025	

8.7 Summary

Methods for analysing special types of contingency tables have been briefly described. Much fuller details are available in the various papers cited. The reader should remember that the techniques for dealing with incomplete tables and for testing for quasi-independence are subject to many technical difficulties not discussed here but which may cause problems in practice. The main purpose of this chapter has been to make readers aware of the existence of such methods, rather than to give a detailed account of their use.

Appendix A

Percentage points of the χ^2 distribution

| d.f. | Probability (P) | | | |
	0.050	0.025	0.010	0.001
1	3.841	5.024	6.635	10.828
2	5.991	7.378	9.210	13.816
3	7.815	9.348	11.345	16.266
4	9.488	11.143	13.277	18.467
5	11.071	12.833	15.086	20.515
6	12.592	14.449	16.812	22.458
7	14.067	16.013	18.475	24.322
8	15.507	17.535	20.090	26.125
9	16.919	19.023	21.666	27.877
10	18.307	20.483	23.209	29.588
11	19.675	21.920	24.725	31.264
12	21.026	23.337	26.217	32.909
13	22.362	24.736	27.688	34.528
14	23.685	26.119	29.141	36.123
15	24.996	27.488	30.578	37.697
16	26.296	28.845	32.000	39.252
17	27.587	30.191	33.409	40.790
18	28.869	31.526	34.805	42.312
19	30.144	32.852	36.191	43.820
20	31.410	34.170	37.566	45.315
21	32.671	35.479	38.932	46.797
22	33.924	36.781	40.289	48.268
23	35.173	38.076	41.638	49.728
24	36.415	39.364	42.980	51.179
25	37.653	40.647	44.314	52.620
26	38.885	41.923	45.642	54.052
27	40.113	43.194	46.963	55.476
28	41.337	44.461	48.278	56.892
29	42.557	45.722	49.588	58.302
30	43.773	46.979	50.892	59.703
40	55.759	59.342	63.691	73.402
50	67.505	71.420	76.154	86.661
60	79.082	83.298	88.379	99.607
80	101.879	106.629	112.329	124.839
100	124.342	129.561	135.807	149.449

Appendix B

Computer software for fitting log-linear and logistic models

The fitting of log-linear and logistic models to complex multi-dimensional contingency tables often involves a considerable degree of computation and would not be feasible on a routine basis without the use of suitable computer programs. Many are available; for example, the three most commonly used statistical packages, SAS, SPSS^X and BMDP, all have comprehensive procedures for categorical data analysis. In SAS, procedure CATMOD is useful for building a wide variety of models for such data. In SPSS^X, the LOGLINEAR program gives maximum likelihood estimates for a range of log-linear and logistic models.

In BMDP programs 4F and LR allow log-linear and logistic models to be estimated and program AR deals with the type of models for ordered data described in Chapter 7. The BMDP program CA provide a simple and flexible means of applying correspondence analysis.

The interactive program GLIM also allows log-linear and logistic models of the type described in this text to be routinely applied. It does, however, also give access to a wide range of alternative models which might be useful in particular situations, for example probit and log-log models.

Other packages which include log-linear modelling and logistic regression are STATSOFT, SYSTAT, and EGRET.

A recent addition to the software available for the analysis of categorical data is STATXACT, which gives **exact** analysis for $r \times c$ contingency tables. In cases where a data set is too large for exact inference, Monte-Carlo methods are used to approximate exact p values and confidence intervals.

References

Agresti, A. (1983). A survey of strategies for modeling cross-classifications having ordinal variables. *J. Am. Statist. Assoc.* **78**, 184–198.

Agresti, A. (1984). *Analysis of Ordinal Categorical Data.* Wiley, New York.

Agresti, A. (1990). *Categorical Data Analysis.* Wiley, New York.

Agresti, A. and Kezouh, A. (1983). Association models for multidimensional cross-classifications of ordinal variables. *Commun. Statist* **A12**, 1261–1276.

Andersen, E.B. (1982). Latent trait models and ability parameter estimation. *Applied Psychological Measurement*, **6**, 445–461.

Armitage, P. and Berry, G. (1987). *Statistical Methods in Medical Research.* Blackwell, Oxford.

Baglivo, J., Oliver, D. and Pagano, M. (1988). Methods for the analysis of contingency tables with large and small cell counts. *J. Am. Statist. Assoc.* **83**, 1006–1013.

Bartlett, M.S. (1935). Contingency table interactions. *Roy. Statist. Soc. Suppl.* **2**, 248–252.

Barnard, G.A. (1984). Discussion of Yates, 1984. *J. Roy. Statist. Soc.* **A147**, 449–450.

Becker, M. and Clogg, C.C. (1989). Analysis of sets of two-way contingency tables using association models. *J. Am. Statist. Soc.* **84**, 142–151.

Bennet, B.M. and Hsu, P. (1960). On the power function of the exact test for the 2×2 contingency table. *Biometrika* **47**, 393–397.

Berkson, J. (1946). Limitations of the application of the fourfold table analysis to hospital data. *Biometrics* **2**, 47–53.

Bhapkar, V.P. (1968). On the analysis of contingency tables with a quantitative response. *Biometrics* **24**, 329–338.

Bhapkar, V.P. and Koch, G.G. (1968). Hypothesis of 'no interaction' in multidimensional contingency tables. *Technometrics* **10**, 107–123.

Birch, M.W. (1963). Maximum likelihood in three-way contingency tables. *J. Roy. Statist. Soc.* (Ser. B) **25**, 220–223.

Birtchnell, J. and Alarcon, J. (1971). Depression and attempted suicide. *Brit. J. Psychiatry* **118**, 289–296.

Bishop, Y.M.M. (1969). Full contingency tables, logits and split contingency tables. *Biometrics* **25**, 383–399.

Bishop, Y.M.M., Fienberg, S.E. and Holland, F.W. (1975). *Discrete Multivariate Analysis.* Massachusetts Institute of Technology Press.

Bock, R.D. (1972). Estimating item parameters and latent ability when responses are scored in two or more nominal categories. *Psychometrika* **37**, 29–51.

Brunden, M.N. (1972). The analysis of non-independent 2×2 tables from $2 \times C$ using rank sums. *Biometrics* **28**, 603–607.

Brunswick, A.F. (1971). Adolescent health, sex and fertility. *Am. J. Public Health* **61**, 711–720.

Chapman, D.G. and Jun-Mo Nam (1968). Asymptotic power of chi-square tests for linear trends in proportions. *Biometrics* **24**, 315–327.

Cochran, W.G. (1954). Some methods for strengthening the common chi-square tests. *Biometrics* **10**, 417–451.

Cohen, J. (1960). A coefficient of agreement for nominal scales. *Educ. Psychol. Measure.* **20**, 37–46.

Connett, J.E., Smith, J.A. and McHugh, R.B. (1987). Sample size and power for pair-matched case-control studies. *Statistics in Medicine,* **6**, 53–59.

Conover, W.J. (1968). Uses and abuses of the continuity correction. *Biometrics* **24**, 1028.

Conover, W.J. (1974). Some reasons for not using the Yates continuity correction on 2×2 contingency tables. *Am. Statist. Assoc.* **69**, 374–376.

Cox, D.R. (1984). Discussion of Yates, 1984. *J. Roy. Statist. Soc.* **A147**, 451.

Cox, D.R. and Hinkley, D.V. (1974). *Theoretical Statistics.* Chapman and Hall, London.

Cox, D.R. and Snell, E.J. (1989). *Analysis of Binary Data,* 2nd Edn. Chapman and Hall, London.

Cramer, H. (1946). *Mathematical Methods for Statistics.* Princeton Univ. Press, Princeton.

D'Agostino, R.B., Chase, W. and Belanger, A. (1988) The appropriateness of some common procedures for testing equality of two independent binomial proportions. *The American Statistician* **42**, 198–202.

Darroch, J.N. (1962). Interactions in multi-factor contingency tables. *J. Roy. Statist. Soc.* (Ser. B) **24**, 251–263.

Darroch, J.N., Lauritzen, S.L. and Speed, T.P. (1980). Markov fields. and log-linear interaction models for contingency tables. *Annals of Statistics* **8**, 552–539.

Davies, C.S. (1991) A one degree of freedom nominal association model for testing independence in two way contingency tables. *Statistics in Medicine*, **10**, 1555–1563.

Deming, W.E. and Stephan, F.F. (1940). On a least squares adjustment of a samples frequency table when the expected marginal totals are known. *Ann. Math. Statist.* **11**, 427–444.

Everitt, B.S. (1968). Moments of the statistics kappa and weighted kappa. *Brit. J. Math. & Statist. Psychol.*, **21**, 97–103.

Everitt, B.S. and Dunn, G. (1991). *Applied Multivariate Data Analysis.* Edward Arnold, Sevenoaks.

Fienberg, S.E. (1969). Preliminary graphical analysis and quasi-independence for a two-way contingency table. *Applied Statistics* **18**, 153–168.

Fienberg, S.E. (1970). The analysis of multidimensional contingency tables. *Ecology* **51**, 419–433.

Fienberg, S.E. (1972). The analysis of incomplete multi-way contingency tables. *Biometrics* **28**, 177–202.

Fienberg, S.E. (1987). *The Analysis of Cross-classified Categorical Data*, 2nd edn. MIT Press, Cambridge.

Fisher, R.A. (1950). *Statistical Methods for Research Workers.* Oliver and Boyd, Edinburgh.

Fleiss, J.L. (1973). *Statistical Methods for Rates and Proportions.* Wiley, New York.

Fleiss, J.L. (1975). Measuring agreement between judges on the presence or absence of a trait. *Biometrics* **31**, 651–659.

Fleiss, J.L. and Cuzick, J. (1979). The reliability of dichotomous judgements: unequal numbers of judgements per subject. *Applied Psychol. Measurement* **3**, 537–542.

Fleiss , J.L. and Everitt, B.S. (1971). Comparing the marginal tools of square contingency tables. *Brit. J. Math. Statist. Psychol.* **24**, 117–123.

Fleiss, J.L., Cohen, J. and Everitt, B.S. (1969). Large sample standard errors of kappa and weighted kappa. *Psychol. Bull.* **72**, 323–327.

Fowlkes, E.B., Freeny, A.E. and Landwehr, J.M. (1988). Evaluating logistic models for large contingency tables. *J. Am. Statist. Assoc.* **83**, 611–620.

Freeman, D.H. (1987). *Applied Categorical Data Analysis*. Marcel Dekker, New York.

Gail, M. and Gart, J. (1973). The determination of sample sizes for use with the exact conditional test in 2×2 comparative trials. *Biometrics* **29**, 441–448.

Gart, J.J. (1969). An exact test for comparing matched proportions in cross-over designs. *Biometrika* **56**, 75–80.

Glass, D.V. (1954). *Social Mobility in Britain*. Free Press, Glencoe, Illinois.

Goodman, L.A. (1968). The analysis of cross-classified data: independence, quasi-independence, and interactions in contingency tables with or without missing data. *J. Am. Statist. Assoc.* **63**, 1091–1131.

Goodman, L.A. (1970). The multivariate analysis of qualitative data: interactions among multiple classifications. *J. Am. Statist Assoc.* **65**, 226–256.

Goodman, L.A. (1971). The analysis of multidimensional contingency tables: Stepwise procedures and direct estimation methods for building models for multiple classifications. *Technometrics* **13**, 33–61.

Goodman, L.A. (1979). Simple models for the analysis of association in cross-classifications having ordered categories. *J Am. Statist. Assoc.* **74**, 537–552.

Goodman, L.A. and Kruskal, W.H. (1954). Measures of association for cross-classifications, Part I. *J. Am. Statist. Assoc.* **49**, 732–764.

Goodman, L.A. and Kruskal, W.H. (1959). Measures of association for cross-classifications, Part II. *J. Am. Statist. Assoc.* **54**, 123–163.

Goodman, L.A. and Kruskal, W.H. (1963). Measures of association for cross-classifications, Part III, Approximate sampling theory. *J. Am. Statist. Assoc.* **58**, 310–364.

Goodman, L.A. and Kruskal, W.H. (1972). Measures of association for cross-classifications, Part IV, Simplification of asymptotic variances. *J. Am. Statist. Assoc.* **67**, 415–421.

Graubard, B.I. and Korn, E.L. (1987). Choice of column scores for testing independence in ordered $2 \times k$ contingency tables. *Biometrics* **43**, 471–476.

Green, P.J. (1984). Iteratively weighted least squares for maximum likelihood estimation and some robust and resistant alternatives (with discussion). *J. Roy Statist. Soc.* **B46**, 149–192.

Greenacre, M.J. (1984). *Theory and Application of Correspondence Analysis.* Academic Press, London.

Grizzle, J.E. and Williams, O. (1972). Log-linear models and tests of independence for contingency tables. *Biometrics* **28**, 137–156.

Grizzle, J.E., Starmer, F. and Koch, G.G. (1969). Analysis of categorical data by linear models. *Biometrics.* **25**, 489–504.

Haberman, S.J. (1973). The analysis of residuals in cross-classified tables. *Biometrics* **29**, 205–220.

Haberman, S.J. (1974). *The Analysis of Frequency Data.* Univ. of Chicago Press, Chicago.

Hays, W.L. (1973). *Statistics for the Social Sciences.* Holt, Reinhart and Winston, New York.

Hills, M. and Armitage P. (1979). The two-period cross-over clinical trial. *Brit. J. Clin. Pharm.* **8**, 7–20.

Hosmer, D.W. and Lemeshow, S. (1989). *Applied Logistic Regression.* John Wiley, New York.

Irwin, J.O. (1949). A note on the subdivision of χ^2 into components. *Biometrika* **36**, 130–134.

Kendall, M.G. and Stuart, A. (1980). *The Advanced Theory of Statistics.* Edward Arnold, London.

Kenward, M. and Jones, B. (1987). The analysis of data from 2×2 crossover designs with baseline measurements. *Statistics in Medicine* **6**, 911–926.

Kimball, A.W. (1954). Short-cut formulae for the exact partition of chi-square in contingency tables. *Biometrics* **10**, 452–458.

Kirk, R.E. (1968). *Experimental Design Procedures for the Behavioural Sciences.* Brooks/Cole Publishing Co., California.

Koch, G.G. and Edwards, S. (1985). Logistic regression. In *Encyclopedia of Statistical Sciences* Vol. 5, pp 128–133. Wiley, New York.

Ku, H.H. and Kullback, S. (1974). Log-linear models in contingency table analysis. *Am. Statistician* **28**, 115–122.

Lancaster, H.O. (1949). The derivation and partition of χ^2 in certain discrete distributions. *Biometrika* **36**, 117–129.

Landis, J.R. and Koch, G.G. (1977). The measurements of observer agreement for categorical data. *Biometrics,* **33**, 159–174.

Larntz, K. (1978). Small-sample comparison of exact levels for chi-squared goodness-of-fit statistics. *J. Am. Statist. Assoc.* **73**, 253–263.

Lee, Sik-Yum and Poon, Wai-Yin (1986). Maximum likelihood estimation of polyserial correlations. *Psychometrika* **51**, 13–121.

Lewis, B.N. (1962). On the analysis of interaction in multidimensional contingency tables. *J. Roy. Statist. Soc.* (Ser. A) **125**, 88–117.

Lewontin, R.C. and Felsenstein, J. (1965). The robustness of homogeneity tests in $2 \times N$ tables. *Biometrics* **21**, 19–33.

Little, R.J.A. (1989). Testing the equality of two independent binomial proportions. *The American Statistician* **43**, 283–288.

McCullagh, P. (1978). A logistic model for paired comparisons with ordered categorical data. *Biometrika* **64**, 449–453.

McCullagh, P. (1980). Regression models for ordinal data (with discussion). *J. Roy. Statist. Soc.* **B42**, 109–142.

McNemar, Q. (1955). *Psychological Statistics.* Wiley, New York.

Mantel, N. (1970). Incomplete contingency tables. *Biometrics* **26**, 291–304.

Mantel, N. (1974). Comment and a suggestion on the Yates continuity correction. *J. Am. Statist. Assoc.* **69**, 378–380.

Mantel, N. and Greenhouse, S.W. (1968). What is the continuity correction? *The American Statistician* **22**, 27–30.

Mantel, N. and Haenszel, W. (1959). Statistical aspects of the analysis of data from retrospective studies of disease. *J. Natl. Cancer Inst.* **22**, 719–748.

Maxwell, A.E. (1970). Comparing the classification of subjects by two independent judges. *Brit. J. Psychiatry.* **116**, 651–655.

Mehta, G.R. and Patel, N.R. (1983). A network algorithm for performing Fisher's exact test in $r \times c$ contingency tables. *J. Am. Statist. Soc.* **78**, 427–434.

Mehta, C.R. and Patel, N.R. (1986a). A hybrid algorithm for Fisher's exact test in unordered $r \times c$ contingency tables. *Communications in Statistics* **15**, 387–403.

Mehta, C.R. and Patel, N.R. (1986b). FEXACT: A Fortran subroutine for Fisher's exact test on unordered $r \times c$ contingency tables. *ACM Trans. Math. Software* **12**, 154–161.

Miettinen, O.S. (1969). Individual matching with multiple controls in the case of all-or-none responses. *Biometrics* **26**, 339–355.

Mood, A.M. and Graybill, F.A. (1963). *Introduction to the Theory of Statistics.* McGraw-Hill, New York.

Olsson, U. Drasgow, F. and Dorans, N.J. (1982). The polyserial correlation coefficient. *Psychometrika* **47**, 337–347.

Payne, C. (1977). The log-linear model for contingency tables. In eds

C.A. O'Muirchcleartagh and C. Payne. *The Analysis of Survey Data*, Vol. 2, Wiley, London.

Pearson, K. (1904). *On the Theory of Contingency and Its Relation to Association and Normal Correlation*. Drapers' Co. Memoirs, Biometric Series No. 1, London.

Pearson, K. (1909). On a new method for determining correlation between a measured character A and a character B, of which only the percentage of cases wherein B exceeds (or falls short of) a given intensity is recorded for each grade of A. *Biometrika* **7**, 96.

Robins, J., Breslow, N. and Greenland, S. (1986). Estimates of the Mantel-Haenszel variance consistent in both sparse-data and large-strata limiting models. *Biometrics* **42**, 311–324.

Rodger, R.S. (1969). Linear hypothesis in $2 \times a$ frequency tables. *Brit. J. Math. Statist. Psychol.* **22**, 29–48.

Roscoe, J.T. and Byars, J.A. (1971). An investigation of the restraints with respect to sample size commonly imposed on the use of the chi-square test. *J. Am. Statist. Assn*, **66**, 755–759.

Roy, S.N. and Kastenbaum, M.A. (1956). On hypothesis of no interaction in a multiway contingency table. *Ann. Math. Statist.* **27**, 749–751.

Schouten, H.J.A. (1985). *'Statistical Measurement of Interobserver Agreement'* Unpublished doctral dissertation. Erasmus University, Rotterdam.

Slakter, M.J. (1966). Comparative validity of the chi-square and two modified chi-square goodness of fit tests for small but equal expected frequencies. *Biometrika* **53**, 619–623.

Somers, R.H. (1962). A new asymmetric measure of association for ordinal variables. *Am. Sociological Rev.* **27**, 799–811.

Somes, G.W. and O'Brian, K.F. (1985). Mantel-Haenszel Statistic. In eds S. Kotz, N.L. Johnson and C.B. Read. *Encyclopedia of Statistical Sciences*, Vol. 5, eds Wiley, New York.

Stuart, A. (1955). A test for homogeneity of the marginal distribution in a two-way classification. *Biometrika* **42**, 412–416.

Taylor, I. and Knowelden, J. (1957). *Principles of Epidemiology.* Churchill, London.

Upton, G.J. (1978). *The Analysis of Cross-Tabulated Data.* Wiley, Chichester.

Ury, H.K. (1975). Efficiency of case-controlled studies with multiple controls per case continuous or dichotomous data. *Biometrics* **31**, 643–650.

Van der Heyden, P.G.M. and de Leeuw J. (1985). Correspondence analysis : a compliment to log-linear analysis. *Psychometrika* **50**, 429–447.

Williams, K. (1976). The failure of Pearson's goodness of fit statistic. *The Statistician* **25**, 49.

Woolson, R.F. and Lachenbruch, P. (1982). Regression analysis of matched case-control data. *Am. J. Epidemiology* **115**, 444–452.

Yates, F. (1934). Contingency tables involving small numbers and the chi-square test. *J. Roy. Statist. Soc. Suppl.* **1**, 217–235.

Yates, F. (1984). Tests of significance for 2×2 contingency tables (with discussion). *J.R. Statist. Soc.* **A147**, 426–463.

Index

Milton Keynes UK
Ingram Content Group UK Ltd.
UKHW020031071024
449327UK00032B/3022